How to Analyze the Cyber Threat from Drones

Background, Analysis Frameworks, and Analysis Tools

KATHARINA LEY BEST, JON SCHMID, SHANE TIERNEY, JALAL AWAN, NAHOM M. BEYENE, MAYNARD A. HOLLIDAY, RAZA KHAN, KAREN LEE

For more information on this publication, visit www.rand.org/t/RR2972

Library of Congress Cataloging-in-Publication Data is available for this publication.
ISBN: 978-1-9774-0287-5

Published by the RAND Corporation, Santa Monica, Calif.

© Copyright 2020 RAND Corporation

RAND® is a registered trademark.

Cover design by Rick Penn-Kraus
Cover images: drone, Kadmy - stock.adobe.com; data, Getty Images.

Support RAND
Make a tax-deductible charitable contribution at
www.rand.org/giving/contribute

www.rand.org

Preface

This report explores the security implications of the rapid growth in unmanned aerial systems (UAS), focusing specifically on current and future vulnerabilities. We propose conceptual approaches meant to enable the enumeration and categorization of UAS-related cyber threats, covering the use of UAS as both targets and vectors of cyberattack, as well as their use by both allies and adversaries. These approaches have been applied to real-world threat scenarios to test their validity and illustrate the types of attacks that are currently feasible. Industry trends and the implications of these trends for cybersecurity are presented as well. Finally, we consider the UAS-related cybersecurity threat from the perspective of the Department of Homeland Security (DHS).

This research should be of interest to individuals within DHS that have responsibilities related to the operation of UAS, those responsible for ensuring the cybersecurity of DHS and DHS-protected facilities and assets, or anyone concerned with the way that UAS proliferation may change the cybersecurity landscape.

About the Homeland Security Research Division

This research was conducted using internal funding generated from operations of the RAND Homeland Security Research Division (HSRD) and within the HSRD Acquisition and Development Program. HSRD conducts research and analysis across the United States Homeland Security Enterprise and serves as the platform by which RAND communicates relevant research from across its units with the broader Homeland Security Enterprise. For more information on the Acquisition and Development Program, see www.rand.org/hsrd or contact Emma Westerman, Director of the Acquisition and Development Program, by email at emma@rand.org or phone at (703) 413-1100 ext. 5660. For more information on this publication, visit www.rand.org/t/RR2972.

Contents

Figures

Tables

Summary

In a world of constant and rapid technological change, minimizing vulnerabilities is a never-ending race against one's adversaries—a race against their technology and its exploitation, as well as their devices, ideas, modes of operation, and tactics. In this report, we examine the cybersecurity implications of one key technological trend: the advancement and proliferation of public-use unmanned aerial systems (UAS). UAS, commonly called drones, have become more common, more readily available, and more sophisticated, supporting new capabilities such as increased data collection and autonomous behavior. As a consequence, UAS are reshaping the cybersecurity world in two key ways. Firstly, UAS present a new kind of critical cybersecurity target. Critical law enforcement or data collection missions using UAS could be undermined by cyberattacks on these platforms. Secondly, UAS in the hands of adversaries could present novel avenues for cyberattacks, with UAS themselves serving as "cyber weapons" intended to deliver malicious content or enable kinetic impacts. In this report, we consider the threat posed by UAS as both targets and vectors of cyberattack. We also examine the relevance of these threats for the Department of Homeland Security (DHS) and DHS-protected facilities and assets.

UAS Security

This work explores the security implications of the rapid growth in UAS technology, focusing specifically on current vulnerabilities and future trends. As drones become more common and sophisticated, both the likelihood and the potential consequences of security threats increase. According to one estimate, sales of piloted or autonomous drones will exceed $12 billion by 2021.[1]

We propose conceptual approaches meant to enable the enumeration and categorization of UAS-related cyber threats. These approaches use both "blue" (ally) and "red" (adversary) mindsets to help policymakers illuminate and understand the kinds of threats that may be facing their organization now and in the future. Blue-mindset-

---

[1] Divya Joshi, "Commercial Unmanned Aerial Vehicle (UAV) Market Analysis – Industry Trends, Companies and What You Should Know," *Business Insider*, August 8, 2017.

based approaches focus on uncovering vulnerabilities in a UAS-related system, while red-mindset-based approaches are built around how to successfully attack a system.

These approaches are applied to real-world threat scenarios to test their validity and illustrate the types of attacks that are currently feasible. These attack examples as well as a structured review of academic and other literature related to UAS-related cyber vulnerabilities help to paint a picture of the likely threat space, today and in the near future.

To enable anticipation of the future threat space related to UAS and cybersecurity, this report also highlights industry trends and the possible implications of these trends. Drones are expected to become more integrated with security and law enforcement functions. New UAS technology developments, including the introduction of more sophisticated, more autonomous control software,[2] and the ability to create drone swarms through mobile networking expand the range and sophistication of potential attacks. UAS controllers and control signals constitute vulnerabilities and points of access for malicious actors. We also explore the use of agent-based modeling techniques to describe threats and help identify robust options for defending against a malicious actor. Agent-based modeling and simulation is a potentially valuable method for understanding drone attacks and can provide meaningful insight for analysts studying how to mitigate potential attacks. Further exploration of these trends in future work could uncover new types of threats and defensive approaches, as well as provide estimates of their expected likelihood and consequences.

UAS Security and the Department of Homeland Security

The findings presented here have implications for DHS. We find DHS to be vulnerable to cyberattacks targeted at DHS-operated UAS (i.e., UAS as cyber targets) and to UAS-enabled cyberattacks (i.e., UAS as cyber weapons).

DHS-Operated UAS as Targets

Four DHS components have documented historical use of UAS in their day-to-day activities: the U.S. Coast Guard (USCG),[3] Customs and Border Protection (CBP),[4] the Federal Emergency Management Agency (FEMA),[5] and the Cybersecurity and Infra-

---

[2] Nick Statt, "Skydio's AI-Powered Autonomous R1 Drone Follows You Around in 4K," *TheVerge*, February 13, 2018.

[3] U.S. Coast Guard, "Unmanned Aircraft System," undated.

[4] U.S. Department of Homeland Security, Customs and Border Protection, "CBP Air and Marine Operations Conducting Third Deployment of UAS at San Angelo," February 27, 2018.

[5] "Drone Use Reaches 'Landmark Level' in Harvey Disaster Response," *InfoGram*, Vol. 17, No. 37, September 14, 2017.

structure Security Agency (CISA).[6] DHS components are using and will continue to use a mix of both DoD-developed (Predator and ScanEagle) and commercially developed UAS. However, with the exception of USCG, all components plan to invest in commercially available UAS going forward. These technologies will make possible or facilitate critical capabilities for each of these components. However, their introduction also means that CBP, FEMA, CISA, and Immigrations and Customs Enforcement (ICE) assets will be vulnerable to the new types of attacks described in this study. That said, given this study's findings regarding the high vulnerability of commercial UAS to cyberattack, components' use of UAS threatens the ability to conduct the following operations at the same time as it provides much-needed new capabilities:

- CBP may lose intelligence, surveillance, and reconnaissance (ISR) capabilities, creating visual blind spots in detection of smuggling or other nefarious activities at borders and ports. CBP may also chose to employ UAS platforms for other activities in the future; for example, *chemical, biological, radiological, nuclear, and explosives* (CBRNE) scanning at ports, where compromised UAS systems could prevent CBP agents from completing their duties, cause significant financial damage by delaying cargo movement while the system is fixed, or even send false "safe" readings of dangerous cargo. Compromised UAS could also create unknowable risk if the CBP operator is unaware of the breach.
- Compromised FEMA UAS may reduce the agency's capability to identify, reach, or supply individuals in peril or medical distress in disaster zones. This may happen both because the compromised UAS asset is no longer capable of performing as intended, or because the UAS could be used to degrade other aerial operations such as helicopter flights or activity of other UAS. Compromised FEMA UAS may also lead to degraded situational awareness if UAS are used for ISR in disaster zones.
- Compromised CISA UAS would degrade the ability of CISA to conduct critical infrastructure inspections in some cases, and could be used in a cyber physical attack to damage the critical infrastructure it was meant to survey. Compromised UAS could also create unknowable risk if the CISA operator is unaware of the breach.
- Finally, ICE intends to use UAS to reduce risk during raids. Compromised ICE UAS would reduce overall capability, require fallback to less-familiar concepts of operation (CONOPS), and increase risk to the agents in the field. Compromised UAS may even create unknowable risk if the ICE operator is unaware of the breach.

---

[6]   U.S. Department of Homeland Security, "Unmanned Aircraft Systems (UAS) - Critical Infrastructure," website, undated.

DHS Attacked with UAS as Cyber Weapons

We find that nearly all DHS components and offices could become victims of a drone-led botnet or data exfiltration attack. These offices and components all have physical locations where sensitive data and wireless networks are prevalent, making them targets for these types of attacks. UAS that have loitering capabilities, for example, those that can land and takeoff again after some period of time, allow this type of covert attack, increasing risk to unhardened systems.

As the ubiquity of connected devices grows, the danger of a drone-injected worm or similar attack, as discussed in Chapter 3, also increases. This attack vector need not be limited to DHS networks and connected devices, because DHS employees' personal devices or home networks could also be access points for nefarious code to gain entry to DHS systems either wirelessly or by an employee connecting an infected device to a DHS laptop.

Recommendations

As a first step in protecting itself from UAS-related cybersecurity attacks or successfully using UAS as cyber assets, DHS can use the approaches outlined in this paper to understand the set of attack vectors and attack surfaces. This is a necessary step, but it is not sufficient for establishing a coherent UAS and cybersecurity plan for cyber defense or offensive cyber operations. Upon gaining a better understanding of the threat space, **DHS must continue to work with senior policymakers, cybersecurity experts, and other government and law enforcement agencies to move toward a coherent UAS cyber strategy**. This work will involve taking inventory of and categorizing UAS platforms, understanding the possible consequences of, as well as mitigation options for, UAS-related cyberattacks, and staying abreast of new technological developments that could change the threat space. DHS should invest in operating a UAS test range (or ranges) in collaboration with the private sector, national labs, and other government stakeholders such as the FAA. This step would help ensure industry compliance with safety and security protocols, and would promote interagency coordination.

DHS should also prioritize the most critical vulnerabilities and find ways to close attack vectors and protect attack surfaces. To understand mitigation options, DHS will need to monitor technological development in counter-UAS (cUAS) systems and experiment with emerging attack techniques and technologies. A coordinated and updateable system of monitoring and intervention is likely to be required as the innovation cycle of cyberattack and countermeasure ensures that even hardened systems cannot be guaranteed immune to attack.

Finally, **DHS will need to monitor UAS adoption and anticipate the implications of widespread UAS diffusion**. Capabilities such as autonomous flight and swarming will widen the UAS application space. As UAS are used in a wider range of activities, the number of legitimate-use UAS that are airborne at any given time will increase. From the perspective of threat mitigation, one of most important tasks in this new UAS-dense environment will be distinguishing licit from illicit activity.

Acknowledgments

We are thankful for the support and guidance of Isaac Porche, John Parmentola, Paul Dreyer, Jordan Fischbach, and Jerry Sollinger, as well as for the substantial contributions from Kenneth Kuhn. We wish to thank the MANA team at New Zealand Defense Technology Agency for providing a reliable agent-model simulation tool. Thanks are also due to Bradley Wilson for providing important technical guidance, Henry Hargrove for advice and counsel, and Daniel Lin for his research contributions to this effort. Daniel Spagiare also provided valuable administrative assistance and proofreading.

Abbreviations

ABM	agent-based modeling
AI	artificial intelligence
ANSP	Air Navigation Service Providers
ATO	air traffic organization
CAD	computer-aided design
CBP	Customs and Border Protection
CISA	Cybersecurity and Infrastructure Security Agency
CBRNE	chemical, biological, radiological, nuclear, explosives
CONOPS	concept of operations
DARPA	Defense Advanced Research Projects Agency
DDOS	distributed denial of service
DHS	Department of Homeland Security
DJI	Dà-Jiāng Innovations
DOS	denial of service
FAA	Federal Aviation Authority
FCC	Federal Communications Commission
FEMA	Federal Emergency Management Agency
FFRDC	federally funded research and development center
GPS	Global Positioning System
HSOAC	Homeland Security Operational Analysis Center
ICE	U.S. Immigration and Customs Enforcement
IMU	inertial measurement units
IoT	Internet of Things
ISR	intelligence, surveillance, and reconnaissance
LTE	Long-Term Evolution
MANA	Map-Aware Non-Uniform Automata

MIT	Massachusetts Institute of Technology
ML	machine learning
NAS	National Airspace System
NASA	National Aeronautics and Space Administration
OFFSet	Offensive Swarm-Enabled Tactics
PEO UAS	Program Executive Office for Unmanned Aerial Systems
RF	radio frequency
S&T	science and technology
SESAR	Single European Sky ATM Research
sUAS	small unmanned aerial systems
SW	software
SWIM	system-wide information management
UAS	unmanned aerial systems
UAV	unmanned aerial vehicles
USCG	U.S. Coast Guard
UTM	UAS Traffic Management

CHAPTER ONE

Introduction

Background and Purpose

In a world of constant and rapid technological change, minimizing vulnerabilities is a never-ending race against one's adversaries—their technology, their devices, their ideas, their modes of operation, their tactics, and their exploitation of technology trends to achieve their political goals. In this report, we examine the cybersecurity implications of one key technological trend: the advancement and proliferation of public-use Unmanned Aerial Systems (UAS). UAS have become more common, more readily available, and more sophisticated, supporting new capabilities such as increased data collection and autonomous behavior. As a consequence, UAS are reshaping the cybersecurity world in two key ways. Firstly, UAS present a new kind of critical cybersecurity target. Critical law enforcement or data collection missions using UAS could be undermined by cyberattacks on these platforms. Secondly, UAS in the hands of adversaries could present novel avenues for cyberattacks, with UAS themselves serving as "cyber weapons" intended to deliver malicious content or kinetic impacts. For instance, UAS swarms carrying explosives in significant numbers can attack U.S. symbols of political power and through cascading effects take down interdependent systems, like critical elements of the U.S. electric power grid.

It can be difficult to predict how emerging technologies translate into new kinds of cybersecurity threats. To help policymakers better understand how UAS are potentially changing the cybersecurity threat space, this report introduces several approaches for inventorying threats related to UAS as cyber targets or cyber weapons. The approaches enable users to identify and categorize threats related to UAS technology, apply a taxonomy of threats to particular scenarios, and visualize the threat space to understand and communicate effectively the nature of the threats and opportunities for improving UAS-related cybersecurity.

When attempting to assess the risks and rewards related to UAS cybersecurity, decisionmakers must approach the subject from several angles. As mentioned above, UAS can serve as both cybersecurity targets and threats to cybersecurity. Additionally, both allies and adversaries could operate UAS under the influence of these two conditions. Figure 1.1 provides some possible examples of threats that fall into each of these four categories, with purple boxes highlighting Department of Homeland Security (DHS) offensive opportunities and blue marking to indicate defensive situations. To capture all of the scenario types described in Figure 1.1, threat enumeration and categorization must include both "blue team" and "red team" mindsets. A blue-team mindset considers how a UAS might be vulnerable or how a system might be vulnerable to UAS-based cyberattacks. A red-team mindset involves devising ways in which a UAS could be attacked or could attack a system.

How This Report Is Organized

In this report, we outline a set of approaches that, from both a blue team and red team perspective, allow for the enumeration and categorization of cybersecurity threats posed by UAS. These approaches are outlined in Chapter Two. (Our analyses do not produce findings on relative likelihood or priority of threats due to consequence modeling or estimation.) In Chapter Three, we use these approaches as the foundation for

Figure 1.1
Categorizing UAS-Related Cyber Threats

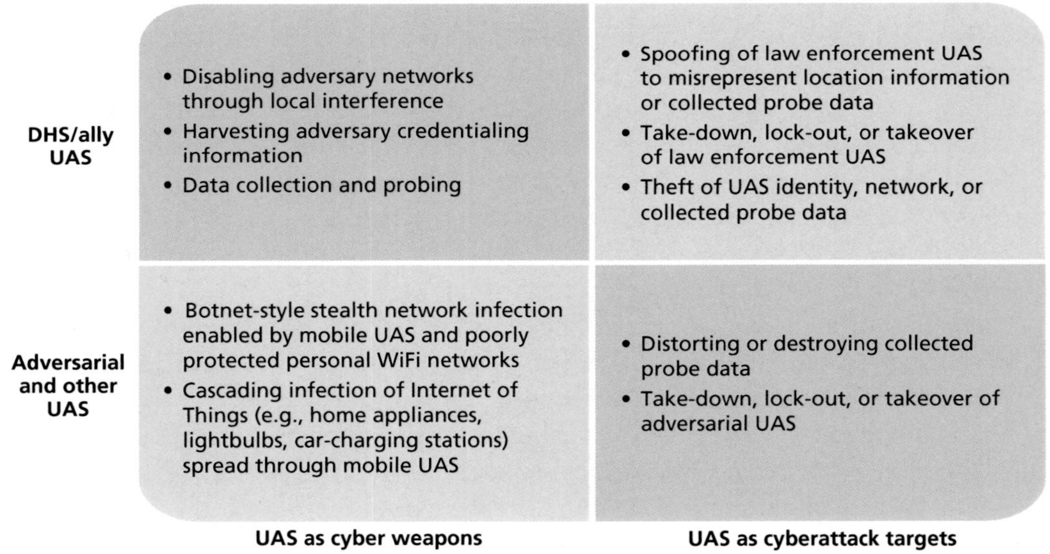

	UAS as cyber weapons	UAS as cyberattack targets
DHS/ally UAS	• Disabling adversary networks through local interference • Harvesting adversary credentialing information • Data collection and probing	• Spoofing of law enforcement UAS to misrepresent location information or collected probe data • Take-down, lock-out, or takeover of law enforcement UAS • Theft of UAS identity, network, or collected probe data
Adversarial and other UAS	• Botnet-style stealth network infection enabled by mobile UAS and poorly protected personal WiFi networks • Cascading infection of Internet of Things (e.g., home appliances, lightbulbs, car-charging stations) spread through mobile UAS	• Distorting or destroying collected probe data • Take-down, lock-out, or takeover of adversarial UAS

a sample review of UAS cybersecurity literature, and we apply the approaches to specific threat scenarios or vignettes. Applying the proposed approaches to specific cases demonstrates the utility of the frameworks in deconstructing attacks and illustrates the range of currently feasible threats. Chapter Four continues the discussion of UAS and cybersecurity into the future, examining how emerging technological trends in UAS development relate to cybersecurity. Trends considered include: the growing use of autonomous (as opposed to remotely piloted) UAS, the development of UAS traffic management systems, "swarming" or group-based autonomous behaviors, the use of machine learning (ML) and artificial intelligence (AI) to detect attacks, increasing hardware complexity and potential technologies for attacking these hardware systems, and the potential use of blockchain technologies. We also explore the use of agent-based modeling (ABM) techniques to describe threats and help identify robust options for defending against a malicious actor. ABM and simulation are potentially valuable methods for understanding drone attacks, and can provide meaningful insights for analysts studying how to best mitigate and thwart potential attacks. Further exploration of these trends in future work could uncover new types of threats and defensive approaches, as well as their expected likelihood and consequences. Finally, in Chapter Five, we consider the UAS cybersecurity threat implications from the perspective of DHS. Specifically, we describe the vulnerability of particular DHS components to the threats described in this report and suggest potential means of threat mitigation. Conclusions and recommendations are offered in Chapter Six. The report also includes an appendix that provides a taxonomy of types of attacks.

CHAPTER TWO

Understanding the UAS Threat Space

Technological progress sometimes has unpredictable consequences. Undesirable applications for new technologies, large costs, and issues related to safety, security, or sustainability can dilute the proposed novelty and benefits promised by technological progress. For example, the widespread use of polychlorinated biphenyl as an insulator and coolant in electronics failed to anticipate the toxic and carcinogenic traits of the compound. Cell phones facilitate communication, yet they can be used to trigger the detonation of improvised explosive devices. While futurists may delight in playing out utopian and dystopian scenarios of technological development as thought experiments, finding ways to uncover risks and benefits of new technologies is also a practical necessity for adapting to a constantly changing world. For example, to prevent an attack on a UAS that is performing a critical law enforcement function, officials must anticipate the various means by which a hacker could gain access to a device, its subcomponents, or its associated software. Similarly, to initiate an effective UAS-enabled cyberattack, government operatives must anticipate the likely cyber defensive measures they will encounter. In this chapter, we introduce three approaches to help planners enumerate and understand potential UAS-related cybersecurity threats by taking the perspective of both defenders and perpetrators of UAS-related cyberattacks.

First, we describe the STRIDE threat model taxonomy as part of our approach for categorizing threats and apply it to UAS as cyber targets and cyber weapons.[1] The STRIDE taxonomy facilitates a "blue team," defensive mindset to help users enumerate possible future threats. STRIDE provides an efficient way of classifying common types of cyber-related threats and encourages practitioners to use this as a framework for brainstorming of potential *attack surface vulnerabilities*. Attack-surface vulnerabilities are the data-transfer points that cannot sufficiently restrict access or privilege of data access, providing points of access that could be exploited in a cyberattack.[2]

---

[1] Adam Shostack, *Threat Modeling: Designing for Security*, Hoboken, N.J.: Wiley, 2014.

[2] Lily Hay Newman, "Hacker Lexicon: What Is an Attack Surface?" *Wired*, March 12, 2017.

Second, we present the cybersecurity kill chain as a means for including the "red team," offensive mindset in threat brainstorming.[3] The cybersecurity kill chain provides a framework within which users can plan possible cyberattacks. Such plans help users take the perspective of the adversary to discover possible attack vectors and weaknesses in the networked communication systems, operating software or applications, and data storage components. *Attack vectors* are the tools, platforms, connections, or security features that could be introduced or exploited to launch a cyberattack.

Finally, we introduce a novel template for capturing the cybersecurity situation in a particular scenario of interest, integrating the attack surface and attack vector approaches to enable a coherent description of cyber-vulnerabilities and opportunities within that scenario. The template also allows for the visual depiction of the set of attack vectors and attack surfaces. The depiction provides an efficient way of rolling up many technical details in support of high-level analysis to identify common attack surfaces and vectors.

Enumerating and Categorizing Threats: The STRIDE Taxonomy

STRIDE Categories of Threats

The first step in protecting against cybersecurity attacks is understanding the possible threat space. For "blue" actors who want to protect against cybersecurity attacks to stay ahead of possible adversaries, they must be creative, proactive, and well informed about adversary capabilities. Enumerating the possible types of future attacks requires a diligent review of the threat space enabled by existing and emerging technologies. It can be helpful for such a review to be rooted in an established framework of possible threat types, and filling in this framework using formalized brainstorming methods to help uncover possible threats can also be helpful. One such framework is Adam Shostack's STRIDE taxonomy for threat modeling, which outlines six areas in which security threats can be classified (and which is outlined in Figure 2.1).[4] Alternatives to STRIDE, such as Gunnar Peterson's DESIST framework, provide other taxonomies for threat enumeration.[5] While the STRIDE taxonomy was originally developed for use in software development, the six areas it covers are also useful for enumerating threats related to cybersecurity and UAS.

The *S* in the STRIDE framework stands for *spoofing* and encompasses the set of threats that violate authentication protocols, enabling an attacker to pretend to be

---

[3] "The Cyber Killchain Framework," website, Lockheed Martin, 2019.

[4] Adam Shostack, *Threat Modeling: Designing for Security*, Hoboken, N.J.: Wiley, 2014.

[5] DESIST refers to dispute, elevation of privilege, spoofing, information disclosure, service denial, and tampering. It was developed by Gunnar Peterson. For description of this model, see Shostack, 2014.

Figure 2.1.
The STRIDE Threat Taxonomy

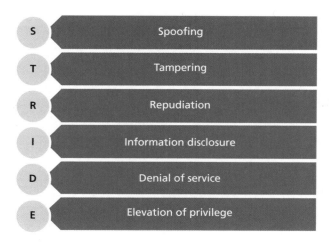

something or someone that he or she is not. In the case of UAS-related cybersecurity, where drones are a target, spoofing could include claiming to be the authorized recipient machine for drone data.

The *T* in the STRIDE framework stands for *tampering*, which involves violating the integrity of a system under attack by making some kind of modification to it. In the case of UAS-related cybersecurity, where drones are used as a cyber weapon, tampering could occur if a drone is used to deliver malware to a target computer using proximity to access an unsecured wireless network. Such malware could potentially infect high-value machinery, such as factory or power plant equipment, or attack such high-impact targets as water systems and power grids.

R stands for *repudiation*, in which attackers refuse to take responsibility for an action. This threat is the least relevant to the domain of UAS-related cybersecurity. One possible example of repudiation is insider abuse of system controls. For example, a drone operator could claim that he or she did not purposefully crash a device by blaming loss of control on a design flaw of the communication network. Another example, where UAS are cyber weapons, could be to distance the identity of an attacker from the consequence by interfering at a communication node loosely affiliated with the point of damage or disruption. This may include using proximity-based network attacks to alter log files of computers that are managing another system that is the target of the attack.

The *I* refers to *information disclosure* and relates to violations of the principle of confidentiality. In information-disclosure attacks, an agent releases information to someone without the proper credentials for receiving it. Information-disclosure threats could include infiltrating a UAS sensor data system to access video, audio, or other data. An agent could also disclose information and later use repudiation to disavow responsibility for this action.

The *D* stands for *denial of service* and refers to denying availability of a resource that is needed for the attacked system to function properly. An example of denial of service is when UAS are targeted, and could involve infecting drone control software to make the devices unresponsive to user inputs.

The last letter of the STRIDE taxonomy, *E*, refers to *elevation of privilege*, which involves violating the principle of authorization to perform an action that one is not allowed to do. An example of authorization of privilege is when UAS are targets, and could involve hijacking of a drone by posing as a legitimate controller. When UAS are used as a cyber weapon, they could be used to deliver data, code, or other signals to debilitate or alter the behavior of the system under attack.[6]

Threat-scenario or vignette development and the STRIDE framework can be combined to support threat brainstorming, model development, testing, and mitigation planning. For the purposes of this report, we focus on STRIDE's ability to support threat brainstorming. The high-level categorization of threats is limited in value for causal analysis but helps aggregate concerns based on a bad actor's strategy for seeking vulnerabilities. Brainstorming can, of course, be performed to create possible threat scenarios. More formal approaches include literature reviews, scenario analysis, analysis of process-flow diagrams, and creation of attack trees. Reviews of historical attacks can also highlight potential vulnerabilities, complementing informal brainstorming. In scenario development and analysis, we use examples of possible adversary actions to understand possible threats.

By working through scenarios of potential future attacks, we can uncover system vulnerabilities. Process- or data-flow diagrams depict the flow of information and data through a system, allowing analysts to identify potential access pathways that an attacker could use. In the case of UAS, where UAS could be a target of attack, such a diagram may depict all of the data connections flowing between UAS and any computing devices assisting in control or data collection. For example, this approach would highlight internal data processing within the drone itself, as well as any wired or wireless connections, information transmission, and processing or data storage infrastructure on external devices. See, for example, Figure 2.2. These diagrams depict not only UAS, but also the controller, other computing devices, and key objects in the environment, such as data-emitting sources encountered on patrol, weather data, computing devices, or beacons for location awareness. The template for capturing cybersecurity threats in a particular scenario that is introduced at the end of this chapter uses such process diagrams.

Attack trees are similar to scenario analysis. An attack tree takes a potential adversary goal (such as "intercept camera feed of drone") and details possible steps that must

---

[6] More details on and examples of these threat types can be found in Chapter 3 of Shostack, 2014.

Figure 2.2
Simple Quadcopter Data-Flow Diagram with Single Radio Controller

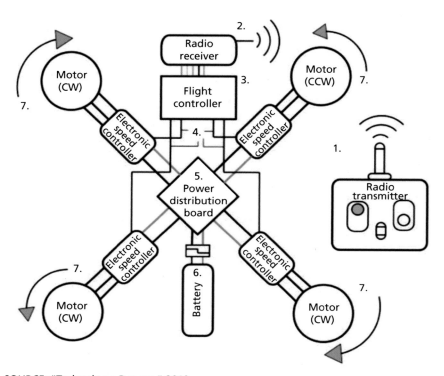

SOURCE: "Technology: Drones," 2019.

be executed to achieve this goal, thus creating a description of a possible attack scenario or vignette. For this hypothetical approach, such steps may include gaining physical access to a computer receiving the data download, intercepting data transmissions in midair, or cracking the password of an online portal that contains the desired data.

After applying some or all of the approaches described above, analysts could populate the STRIDE framework (or a similar one) with the set of threat scenarios to be considered when making cybersecurity decisions. This defensive blue team approach identifies the attack surfaces that adversaries could exploit, and it starts identifying potential loopholes that may need to be closed or mitigated to strengthen cybersecurity of a UAS-related system.

Discovering Threats in a Scenario: The Cybersecurity Kill Chain

Within a given attack scenario, great diversity exists in how a drone may be found to be vulnerable. Forcing a decisionmaker to take the perspective of an adversary (i.e., to play the red team), may help identify threats that were not uncovered by the "blue perspective" STRIDE framework method described in the last section. Supporting this red team approach, the cybersecurity kill chain enables a user to identify when and how a particular system is vulnerable within a scenario. This could enable the design of an informed defense against a specific threat. For example, such an approach could identify a weak link in a long chain of communications, such as failure to secure a wireless signal at an employee's home, rather than focusing efforts on further strengthening encryption at the central data warehouse.

The cybersecurity kill chain identifies seven stages—reconnaissance, weaponization, delivery, exploitation, installation, command and control, and actions—of a cyberattack. The chain, depicted in Figure 2.3, represents an ordered sequence in which each stage represents an action taken by an adversary. Critically, each stage also presents an opportunity for attack detection. As the stages are sequential, early detection is associated with less-disruptive consequences and less-costly fixes. Because the appropriate defensive action depends on where in the chain a given action is located, specifying where the drone sits on the cybersecurity kill chain facilitates the adoption of effective security measures.

In many attack scenarios involving UAS, the purpose of initiating the cybersecurity kill chain is to take an action against the drone itself. Such attacks seek to gain control over a drone or its subsystems to, *inter alia*, capture or alter its data, change its course, or destroy the device. Within such attacks, the drone plays a role in every segment of the cybersecurity kill chain. We call this variant of UAS-enabled cyberattacks "UAS as target."

In other UAS-enabled attacks, aggressors exploit the unique characteristics of UAS as a means to attack a different (non-UAS) target. In such cases, the drone is used in at least one intermediate link of the cybersecurity kill chain to take action on some other target, or in the final link to facilitate the action. While such attacks may exploit the same security vulnerabilities that are used to target UAS directly, the UAS is used as means to an end in the latter category of attack. We refer to such attacks as "UAS as vector." The utility of the "UAS as target" versus "UAS as vector" distinction is that

Figure 2.3
Cybersecurity Kill Chain

different attack types are associated with different defense postures. In the conclusion of this report, we briefly elaborate on appropriate defense approaches to each attack variant.

Visualizing Threats: The UAS Cybersecurity Diagram Template

Using both blue- and red-perspective approaches to enumerate possible vulnerabilities can reveal a complex and intimidating threat space. In this section, we introduce a novel UAS cybersecurity diagram template intended to portray the threat space in an easy-to-understand way. Applications of the diagram template help to clarify where and how hacks are likely to pose a threat by providing a visual way to capture where a system may be vulnerable to attack (*attack surfaces*) and how an attack can access the system (*attack vectors*). The visual depictions can also help communicate likely types of threats effectively for a broad range of audiences. The template captures attack surfaces and attack vectors separately, resulting in two complementary illustrations. In the next chapter, we apply this template to several vignettes and provide concrete examples of its application.

Attack-Surface Illustration

The attack-surface illustration template includes three core nodes that define the boundaries for communications, as shown in Figure 2.4. These nodes are the human operator, the drone itself, and the drone environment, meaning that communication channels exist between the human operator and the drone as well as between the drone and the environment. Additional detail can be captured by creating a node for the human's environment when a UAS application involves operation beyond line of sight.

The communication channel between the human operator and the drone handles operation and control of the drone. The first link in this system is a person, who can be viewed as the primary operator of the drone, but the transmission of human commands to the drone may be delivered across a variety of computing devices (e.g., charg-

Figure 2.4
UAS Attack-Surface Illustration Template

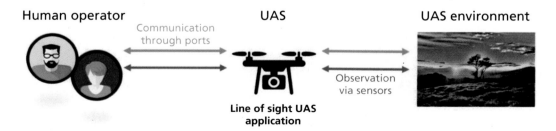

Figure 2.5
UAS Information from Communications and Observations

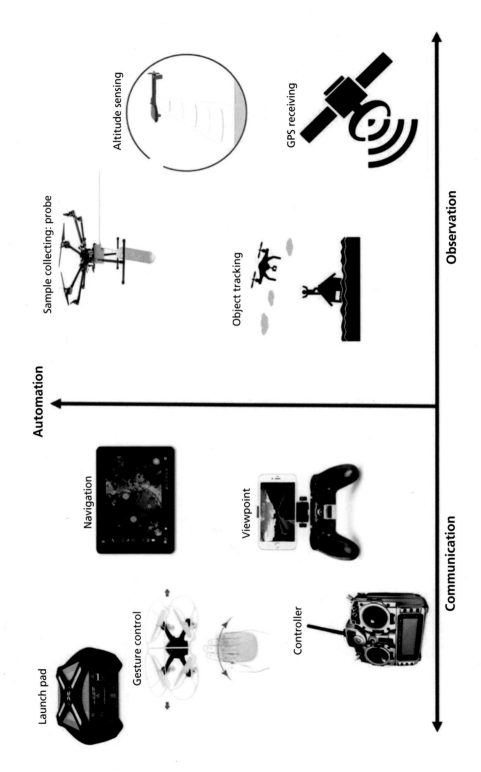

ing base with flight schedule instructions, cloud-based server for in-flight commands, physical controller, cell phone, laptop, tablet), all of which can be highlighted in the image template to point out potential attack surfaces. The left-hand side of Figure 2.5 provides some examples of additional nodes, such as navigation software or controllers, that could be added to the image template to show possible points of attack.

The communication channel between the drone and its operational environment is focused less on communications needed for UAS control and more on information gathered from the operational environment. The collection of sensory information by the drone can involve a variety of sources, such as those shown on the right-hand side of Figure 2.5, to pick up a variety of environmental factors such as GPS signal, altitude information, and visual or other sensor data. The template can be annotated with these nodes.

Attack-Vector Illustration

In conjunction with the attack-surface illustration template, we present a template for illustrating the vectors that could be used in a cyberattack (Figure 2.6). The attack-vector illustration answers the following five questions about the attack being illustrated:

1. What was the bug (cyber weapon)?
2. How did it get there (attack vector)?
3. Where did it get in (system access point or attack surface)?
4. What failed (security vulnerability or candidate for mitigation)?
5. What happened (consequence)?

To be able to depict attack vectors visually across the UAS and cybersecurity threat space, the attack-vector template is intended to isolate three sequential segments of the cybersecurity kill chain (see Figure 2.3) that align with three key phases of a full-scale cyberattack. In the first type of sequence, we show the process of planning or infiltration by detailing activities associated with Reconnaissance and Weaponiza-

Figure 2.6
UAS Attack-Vector Illustration Template

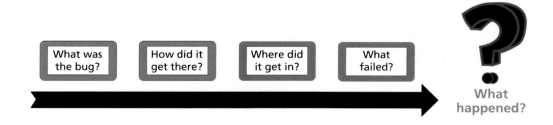

tion. The template focuses on the act of intelligence-gathering and formulation of the cyberattack. The next type of sequence concerns the process of modifying the system through activities of delivery, exploitation, and installation. In this sequence of illustration, the delivery of a weapon and violation of privilege identification are shown with the result of altering network or functional settings. Lastly, a sequence regarding the hijacking of UAS operations presents activities associated with command and control and actions. This segment accounts for the threat conditions regarding damages, harms, degredation, disruption, and denial involving UAS applications and operations.

Integrating the visualization templates developed above for attack surfaces and attack vectors, we can use a "split-screen" approach to provide a complete visual representation of a cyberattack related to UAS. An example of such an image is shown in Figure 2.7, and this method is used to illustrate the vignettes included in Chapter Three.

The methods introduced in this chapter provide a means for enumerating, assessing, and depicting UAS-related cybersecurity threats. Together, these methods

Figure 2.7
Example of a "Split-Screen" Illustration for Attack Surface and Attack Vector

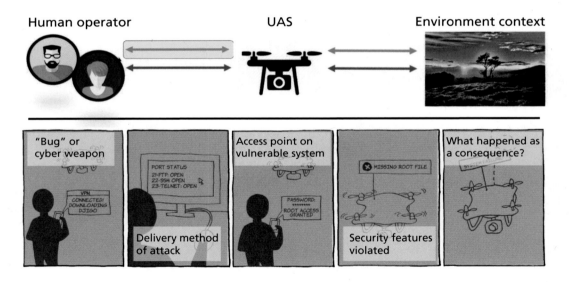

offer both an ally and adversary perspective for the identification of threats, as well as a visual method for understanding attack vectors and surfaces. In the next chapter, these methods are applied to specific cyber-vulnerability vignettes to demonstrate the usefulness of these methods and provide an understanding of the vignettes themselves.

CHAPTER THREE

The UAS and Cybersecurity Threat Space Today

Vulnerabilities of commercially available UAS as well as cybersecurity attacks that use UAS are well documented in the academic literature, popular media, and nontraditional media sources such as social media and blogs. In this chapter, we survey these sources to describe the current UAS and cybersecurity threat space. This chapter first considers the array of plausible threats by surveying the literature on documented UAS exploits. By surveying and coding (e.g., method, target, actor) the existing literature on the topic, we are able to create an inventory of attack types and make preliminary judgments regarding their relative prevalence. However, while the variety and sheer number of documented exploits provide clear evidence of the overall cyber vulnerability of UAS, deriving operationally relevant information from these data requires the application of a conceptual framework. Therefore, following a description of the threat space, we apply the approaches described in Chapter Two to four selected cases of UAS-related cybersecurity exploits to illustrate how policymakers can make the leap from simply inventorying threats to developing plans for cyber defense.

The Extent of UAS Cyber Vulnerabilities

Sander Walters has compiled a list of 26 instances of documented UAS exploitation.[1] Careful scrutiny of the method of attack, the target drone, and the individuals responsible for the exploit reveals a large aggregated attack surface, a wide range of sophistication in terms of the particular UAS that was exploited, and a low threshold in terms of the required computational competence of the adversary. In regard to the overall attack surface, examining the documented exploits reveals many distinct vulnerabilities spread across all of the primary UAS subsystems. For example, successful exploits targeted poor passphrase security, known default settings, and unprotected ad-hoc networks. In term of subsystems, vulnerabilities were exploited in the UAS themselves as

---

[1] Sander Walters, "How Can Drones Be Hacked? The Updated List of Vulnerable Drones and Attack Tools," webpage, October 29, 2016.

well as their receivers, optical sensors, controllers, navigation apps, and all of the communications links connecting these subsystems.

In regard to the required technological skill of the adversary, we found that most UAS-enabled exploits do not require a high degree of sophistication. One exploit took control of a UAS using a Raspberry Pi: a rudimentary and inexpensive ($35) computer intended to teach basic computer competency. The fact that the means to conduct a UAS exploit are publicly available lowers the threshold for adversary competence. In many cases, the individuals responsible for the attack document their methodology on websites such as YouTube or personal blogs. Indeed, in many cases, the code used in the exploit is posted to such searchable code repositories as GitHub.

Nor are these exploits limited to early-generation or low-end UAS.[2] Documented exploits have targeted the DJI Phantom 4, valued at $1,500; the DJI Inspire, valued between $2,000 and $3,000; and the Yuneec Tornado, valued at $3,000.[3] Similarly, high-grade controllers such as the FrSky ACCST and the DJI Naza-M controller have been successfully exploited. One IT security consultant even hijacked a professional-grade Aerialtronics Altura Zenith UAV, valued between $25,000 and $35,000, that is used in law enforcement.[4]

Additional evidence regarding the vulnerability of UAS is found in a series of sophisticated demonstrations undertaken by university-based researchers. For example, research by Junia Valente, a PhD Candidate at the University of Texas at Dallas, led the U.S. Computer Emergency Readiness Team (US-CERT) in 2017 to issue a Vulnerability Note on a family of quadcopters that Valente demonstrated could be anonymously hijacked through their local FTP network. In fact, university-based researchers have demonstrated the vulnerability of commercially available UAS to a wide range of attacks, including internet-facing botnet attacks,[5] ad-hoc network attacks,[6] data collection and probing,[7] UAS location detection and tracking,[8] UAS hijacking or take-

---

[2]   Walters, 2016.

[3]   UAS cost data come from Association for Unmanned Vehicle Systems International, "Unmanned Systems and Robotics Database," webpage, undated.

[4]   Association for Unmanned Vehicle Systems International, undated.

[5]   Theodore Reed, Joseph Geis, and Sven Dietrich, "SkyNET: A 3G-Enabled Mobile Attack Drone and Stealth Botmaster," *Proceedings of the 5th USENIX Conference on Offensive Technologies*, San Francisco, Calif.; WOOT '11, 2011.

[6]   Reed et al., 2011; Eyal Ronen, Adi Shamir, Achi-Or Weingarten, and Colin O'Flynn, "IoT Goes Nuclear: Creating a ZigBee Chain Reaction," *2017 IEEE Symposium on Security and Privacy*, San Jose, Calif.: IEEE, June 2017.

[7]   Reed et al., 2011.

[8]   Andrew J. Kerns, Daniel P. Shepard, Jahshan A. Vhatti, and Todd E. Humphreys, "Unmanned Aircraft Capture and Control Via GPS Spoofing," *Journal of Field Robotics*, Vol. 31., No. 4, July/August 2014; Fernando Trujano, Benjamin Chan, Greg Beams, and Reece Rivera, "Security Analysis of DJI Phantom 3 Standard," Massachusetts Institute of Technology, May 11, 2016; and Junia Valente and Alvaro E. Cardenas, "Understanding

down,[9] media capture,[10] and the modification of software to allow UAS entry of FCC-prohibited airspace.[11]

To understand the types of threats that are prevalent in the literature, we aggregate citations of historical or possible future attacks across a variety of sources and categorize them based on attack type, UAS role (as target or as cyber weapon), and access points (attack surface, attack vector, and type of bug or weapon used). A complete table to categorize attacks is provided in Appendix A. Overall we find, through the STRIDE taxonomy, that most of the cyberattacks documented across different types of sources use either denial-of-service or spoofing attacks to hijack active UAS. These attacks target UAS open networks, such as WiFi or RC connections, and use radio frequencies to overpower the original owner's signal. Attackers then seek to replace the displaced signal with their own, delivering new instructions to the drone. In the majority of examples from the literature, UAS are the targets of cyberattacks, with UAS rarely used as cyber weapons. In the cases where UAS are used as cyber weapons, they are generally modified to include a network or radio frequency scanner, which is the true offensive cyber weapon used in the attack. Figure 3.1 provides a breakdown of the frequency of different attack types, UAS role, and attack surfaces exhibited by the examples outlined in Appendix A.

Vulnerabilities in public-use UAS are ubiquitous. In the next section, we review the literature of documented UAS exploits across sources. Since merely documenting this fact does little to suggest appropriate cyber-defense strategies, we follow this literature review with a more detailed survey of four particular vignettes. In these vignettes, we apply the approaches described in Chapter Two to move toward operationally relevant insight into how to defend against the cyber UAS threat.

Vignettes of Selected UAS Cyberattacks

The vignettes that follow describe in greater detail four of the most sophisticated demonstration attacks on commercially available UAS. The first two describe instances where UAS have been targeted directly. The second two vignettes describe cases in which researchers have used a drone to gain proximity to a target, gather data, and then deliver malware through the drone's ad-hoc network. For each case, we point out how the STRIDE taxonomy along with the cybersecurity kill chain afford analytical

---

Security Threats in Consumer Drones Through the Lens of the Discovery Quadcopter Family," *Proceedings of the 2017 Workshop on Internet of Things Security and Privacy*, Dallas, Tex.: Association for Computing Machinery, 2017.

[9]   Kerns et al., 2014; Trujano et al., 2016; Valente and Cardenas, 2017.

[10]   Kerns et al., 2014; Trujano et al., 2016; Valente and Cardenas, 2017.

[11]   Trujano et al., 2016.

leverage to the task of understanding the exploit. In particular, the STRIDE framework is useful in classifying the attack methodology, and the cybersecurity kill chain allows for the segmentation of the attack into discrete defensible stages.

UAS as Targets One: Hijacking a Drone Remotely

Event Summary
"We were able to access and delete files at the root while the drone was midflight. If an attacker were to delete the entire file system the drone would most likely crash."[12]

Description of Attack
Four researchers from the Massachusetts Institute of Technology (MIT) spent a month identifying and exploiting security vulnerabilities on a popular drone (the DJI Phantom 3 Standard). The researchers used network-mapping tools to capture outgoing packets from the drone's three major subsystems: the drone, its camera, and its controller. Once each subsystem was identified, researchers gained root access by exploiting poor device password security. Root access to the drone's file system allows for the modification of system files, which, in turn, allows an attacker to modify flight paths

Figure 3.1
Breakdown of Prevalence of Selected Cyberattack Characteristics in Literature

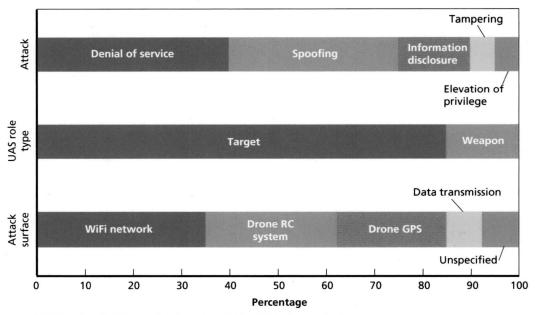

SOURCE: The data in this graph is based on Table A.1 in Appendix A.

---

[12] Trujano et al., 2016, p. 6.

or crash the device. Root access to the camera would allow attackers to access, delete, or add images or video. Finally, the researchers identified a vulnerability in the drone's Android app that would allow an attacker to bypass software-imposed restrictions on entering Federal Communications Commission (FCC)-prohibited airspace.

Breaking Down the Threat

The STRIDE framework is useful in classifying the types of threats illustrated by this case. Researchers were able to gain root access to all of the drone's major systems with relative ease (elevation of privilege). Root access allowed for the modification of system files (tampering). Researchers gained access to the drone's WiFi network and file system through the use of publicly known default passwords (information disclosure).

Applying the cybersecurity kill chain to the attack indicates that the attack could have been prevented by preventing the reconnaissance stage of the chain. The majority of public-use UAS create ad-hoc networks to link the device with their controllers. The attack in question demonstrates the continued vulnerability of these networks to the application of network mapping and discovery tools (i.e., their vulnerability to reconnaissance).

Figure 3.2 provides the attack surface and attack vector visualization as described in Chapter Two for this vignette.

Figure 3.2
UAS Attack to Access and Delete Files Midflight

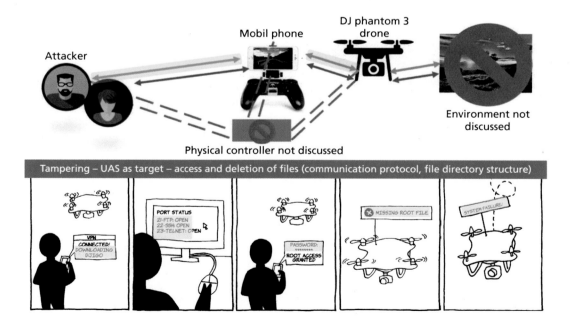

UAS as Target Two: GPS Spoofing

Event Summary

"A field test showed that a destructive GPS spoofing attack against a rotorcraft UAS is both technically and operationally feasible."[13]

Description of Attack

Four researchers at the University of Texas at Austin proposed and implemented a method to gain control over a public-use drone by the transmission of a deceptive GPS signal (i.e., GPS spoofing). In the proposed attack, a spoofing device first receives legitimate signals from GPS satellites. The spoofer then generates a series of counterfeit signals that force the drone receiver to transmit phantom position and velocity signals. Once the spoofer has exerted control over the device, the spoofer can manipulate the drone's flight path or crash the vehicle entirely.

Breaking Down the Threat

Within the STRIDE framework, this attack constitutes a spoofing attack. To navigate reliably, UAS typically combine information from an internal inertial measurement units (IMU) and GPS satellites. Civil GPS signals are characterized by few security measures. Gaining control of the GPS information received by a drone allows the manipulation of the drone estimates of position and velocity. Manipulation of these estimates enables an attacker to hijack or crash the target drone. GPS spoofing can be undertaken covertly and from considerable distance.

Applying the cybersecurity kill chain framework provides insight about how the attack could have been prevented. In this case, weaponization (i.e., the configuration and acquisition of a spoofing device) is almost undetectable to the victim. In contrast, the delivery stage (when the UAS accepted the counterfeit signals) could have been prevented by anti-spoofing defense measures such as distortion detection and direction-of-arrival sensing. The presence of these mitigation features would support a sustained feedback loop to the human operator to apply alternative control strategies or features until the primary control strategy could be restored with confidence.

Figure 3.3 provides the attack surface and attack vector visualization as described in Chapter Two for this vignette.

---

[13] Kerns et al., 2014, p. 26.

Figure 3.3
UAS Attack to Fool Hovering Feature with Spoofed GPS Signal

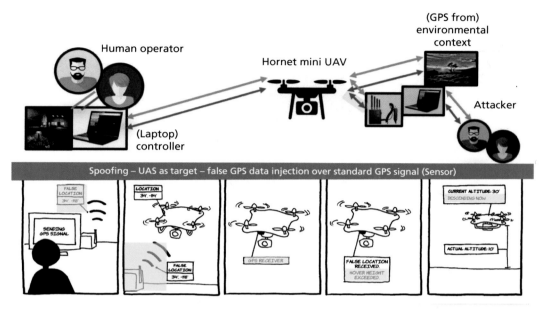

UAS as Vector One: A Drone Botmaster

Event Summary

"[T]ake advantage of poorly configured wireless network security, and poor trust configurations on mobile devices, to join networks and access devices locally using a mobile attack drone[.]"[14]

Description of Attack

Three researchers from Stevens Institute of Technology propose a method to use one drone first to build and then control a hidden internet-facing botnet. In the proposed attack, an enhanced drone makes three flights over an urban area. During the first flight, the drone surveys and collects information on the WiFi networks within the area of attack. The second flight is used to access vulnerable networks. During the final trip, the drone joins the compromised networks and enlists local hosts into a botnet.

Breaking Down the Threat

Within the STRIDE framework, the attack in question constitutes a DOS threat. A botnet refers to a network of malware-compromised devices that can be used to attack network-connected devices. Botnets represent a major cybersecurity risk. They can be used to execute distributed-denial-of-service (DDOS) attacks, steal data, and

---

[14] Reed et al., 2011, p. 2.

hijack devices. Botnets are controlled by a botmaster. In the proposed attack, the use of drones allows for control of a botnet in a way that hides the botmaster. Thus, the primary threat implied in the proposed attack methodology is the potential to use commercially available drones to anonymously initiate cyberattacks via botnets.

Applying the cybersecurity kill chain framework to this case is useful for determining the role of the drone in the larger attack. Figure 3.4 depicts where the drone sits on the cybersecurity kill chain for this particular attack.

The figure demonstrates that the role of the UAS is instrumental; it is used to gain proximity and surveil local networks (reconnaissance), deliver malware (delivery), and anonymously control a botnet. Figure 3.5 provides the attack-surface and attack-vector visualization, as described in Chapter Two, for this vignette.

Figure 3.4
UAS and the Cybersecurity Kill Chain—"UAS as Vector" Exploit (A Drone Botmaster)

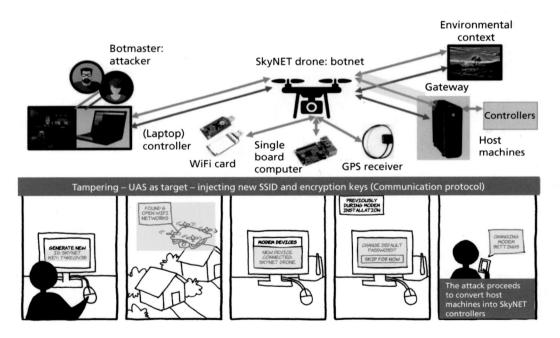

Figure 3.5
Attack by Drone Hijacks Open Networks and Overrides Networked Devices

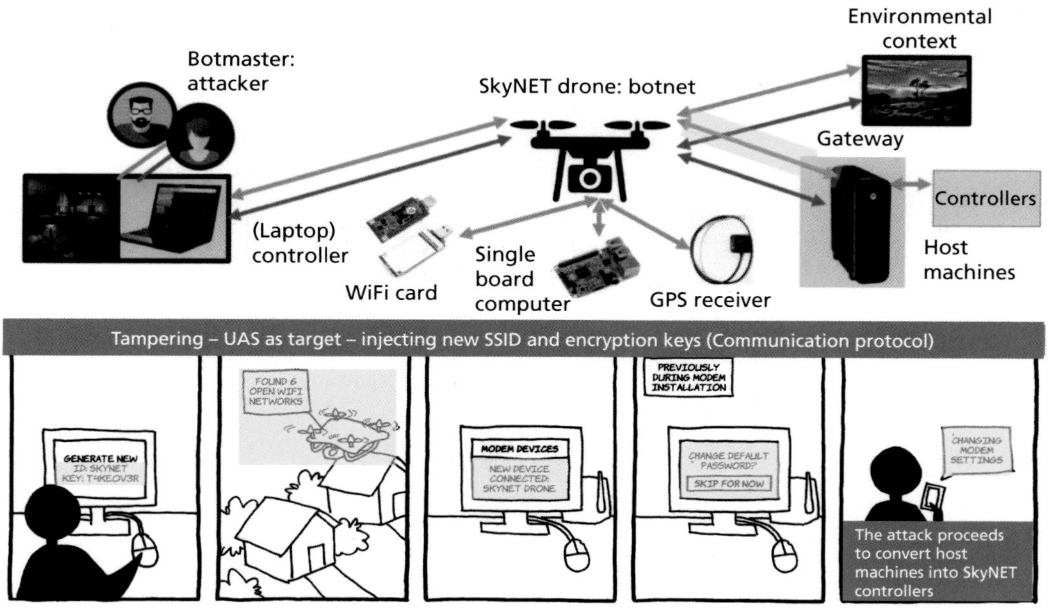

UAS as Vector Two: A Drone-Injected Worm

Event Summary

"We show that without giving it much thought, we are going to populate our homes, offices, and neighborhoods with a dense network of billions of tiny transmitters and receivers that have ad-hoc networking capabilities[.]"[15]

"By using this new communication medium to spread infectious malware from one IoT device to all its physically adjacent neighbors, hackers can rapidly cause city-wide disruptions which are very difficult to stop and to investigate[.]"[16]

Description of Attack

Four university researchers based in Israel and Canada used a DJI drone to inject a worm and take control of smart lightbulbs in an office building in Be'er Sheva, Israel. The attack exploited a security flaw in the Zigbee communications protocol that is used to connect the bulbs. The researchers used the drone to arrive sufficiently close to the bulbs to issue a factory reset command. The drone's software then updated the devices' firmware, took control of the bulbs, and made them blink "SOS" in Morse code.

---

[15] Ronen et al., 2017, p. 1.

[16] Ronen et al., 2017, p. 1.

Breaking Down the Threat

Within the STRIDE framework, the attack constitutes a case of tampering, as the attack injected malicious code that modified the software of the smart lightbulbs, allowing them to be controlled remotely by the researchers.

Again, applying the cybersecurity kill chain framework to the attack reveals the stages during which the drone was used in the attack. In this case, the drone was used during the delivery and the command and control stage of the cybersecurity kill chain. More generally, it appears that in cases in which UAS are used as attack vectors, the drone enters the cybersecurity kill chain to exploit the particular advantages afforded by UAS (typically the ability to get close to a target or to create and access insecure networks). Executing the delivery stage of the cybersecurity kill chain via the Zigbee communications protocol is particularly noteworthy. The Zigbee protocol is a common communication protocol for Internet of Things (IoT) devices. While the attack in question was merely meant to indicate the presence of a security flaw and was thus benign, the attack methodology could be used to disable connected devices permanently or to initiate a DDOS attack. Further, by using the Zigbee wireless communication to spread the worm, the attack avoids the security measures and traffic monitoring associated with internet communication.

Figure 3.6 provides the attack-surface and attack-vector visualization, as described in Chapter Two, for this vignette.

Figure 3.6
Attack by Drone to Overcome Proximity-Based Control of Smart Lightbulbs

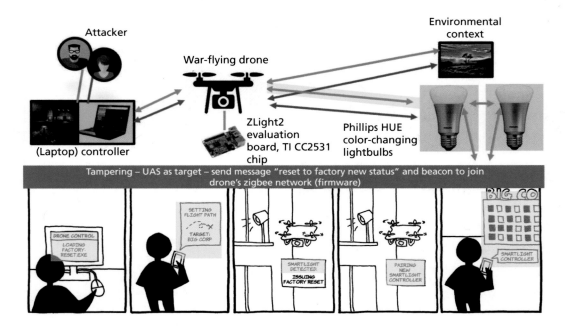

In this chapter, we explored the current literature on UAS and cybersecurity both at a high level and by applying the frameworks from Chapter Two to specific vignettes. Both approaches provide a starting point for enumerating possible threats and targets in a more specific context. In the next chapter, we explore how specific technological trends may change this threat space in the future.

CHAPTER FOUR

Industry Trends and the Future of UAS Cybersecurity

The pace of technological change in UAS technology exacerbates the cyber threat posed by UAS. In 2017, 7,356 patents were filed for UAS-related innovations. Of this total, 5,696 (77 percent) were assigned to Chinese organizations. When technology changes at such a clip, cybersecurity professionals are often left playing catch-up. Given concerns about the security of Chinese-manufactured UAS,[1] the apparent dominance of Chinese firms in this sector should give U.S.-based cybersecurity professionals pause. Although patent activity is an imperfect proxy for technological innovation, large shifts in global patent production are useful in identifying national science and technology (S&T) priorities and secular trends in the distribution of innovation for a given technological domain.

Besides the overall rate of UAS-related technological innovation, emerging industry trends may intensify threats described in earlier chapters of this report. The trends that advance the capabilities of UAS can modulate the perceived benefit by exposing users and bystanders to risks of unauthorized elevation of privilege or violation of information assurance, which were further detailed by the STRIDE framework. At the same time, these trends have the potential to mitigate the threat. For example, as UAS are equipped with additional autonomous flight capabilities and human operators become less common, the chances of aberrant system behavior going unnoticed may increase, particularly if automated detection systems are not deployed. On the other hand, automated tools could be used to identify and respond to attacks. Other technological trends relevant to the overall cyber threat of UAS include the growing sophistication and use of autonomous flight capabilities, UAS traffic management systems, "swarming," the use of ML and AI to detect cyber intrusions, and the use of blockchain technologies to log data and ensure secure communications. The section that follows first examines the overall pace of technological change in UAS technology, and then examines the particular emerging trends mentioned above.

---

[1] Paul Mozur, "Drone Maker D.J.I. May be Sending Data to China, U.S. Officials Say," *New York Times*, November 29, 2017.

Technological Innovation and UAS

Patents are a common proxy for technological innovation. A patent is a property right on an innovation that gives its owner the exclusive right to use, transfer, or contract for the underlying innovation. To attain a patent, an applicant must demonstrate to a patent examiner with subject matter expertise in the relevant technological domain that the underlying innovation is nonobvious, novel, and useful. This condition assures that the innovations underlying patents refer to improvements from the status quo. This is not to say that all patents protect equally important inventions, or even that trivial innovations are not sometimes patented. However, for the most part, patents are a useful means of measuring rates of technological change especially when used in large aggregates such as they are here.

Figure 4.1 shows the annual number of UAS patents, as well as the cumulative output. The annual levels illustrate the rapid growth rate of UAS patenting over time. Since the first UAS-related patent was granted in 1981, 19,333 patents have been granted for technologies related to UAS.[2] The time trends illustrated in Figure 4.1 indicate that UAS patenting is currently growing at an exponential rate. From 2015

Figure 4.1
Number of UAS Patents

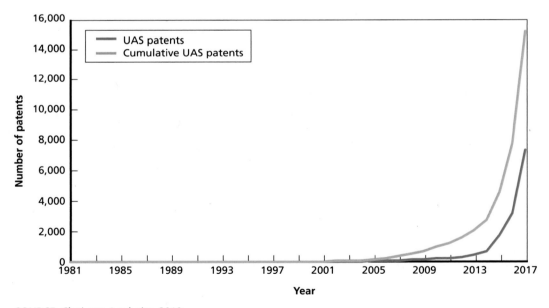

SOURCE: Clarivate Analytics, 2019.

---

[2] Results of the following query of the Derwent Innovations Index (Clarivate Analytics, 2019): "TS=(UAV AND unmanned) OR TS=(UAS AND unmanned) OR TS=("unmanned aerial") OR TS=(drone AND unmanned)."

to 2016, the rate of patent growth was 76 percent. From 2016 to 2017, UAS patenting grew at a rate of 130 percent.

The cumulative curve is useful in assessing where UAS innovation presently sits along a technology innovation S-curve. Rogers (2003) finds that technology adoption typically follows a predictable pattern.[3] Specifically, rates of innovation typically follow a bell-shaped curve. Plotting cumulative adoption or cumulative innovation over time results in an S-shaped or logistic curve. Figure 4.2 provides a standard S-curve. While technological forecasting is beyond the scope of this research, combining Rogers' insight with the observation that UAS patenting is in the exponential portion of its S-curve suggests that UAS patenting will likely continue to be high in the near term.

A closer look at the sources of this growth suggests that recent growth has been driven largely by patenting by Chinese organizations. Figure 4.3 depicts annual patent output for organizations from the United States, China, and the rest of the world. From 2007 to 2017,[4] China's average annual rate of patent growth was 191 percent.

Figure 4.2
Generic S-Curve

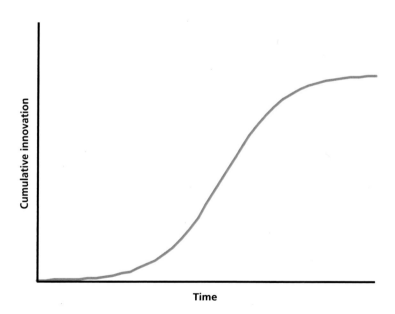

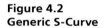

Time

Text results were then cleaned, parsed, and analyzed using VantagePoint.

[3]   Everett M. Rogers, *Diffusion of Innovations*, 5th ed., New York: Free Press (Simon and Schuster), 2003, p. 576.

[4]   Clarivate Analytics, "Derwent Innovations Index," database, 2019.

Figure 4.3
Annual Patent Output for the United States, China, and the Rest of the World, 2001–2017

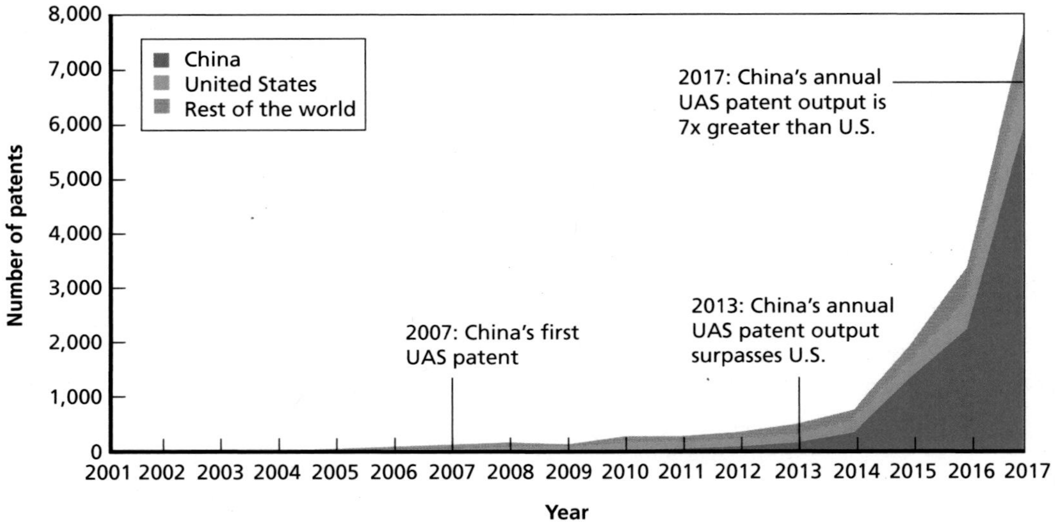

SOURCE: Clarivate Analytics, 2019.

The average growth rate for the United States and the rest of the world was 30 percent and 40 percent, respectively.[5]

Considering the organizations responsible for UAS patenting is also illustrative. Table 4.1 lists the top UAS patenting organizations. The table indicates that Chinese and U.S. organizations are responsible for the lion's share of UAS innovation. It is interesting to note that U.S. innovation in UAS is predominantly driven by firms, while Chinese UAS innovation is more evenly divided between the private sector, universities, government labs, and state-owned enterprises.

Emerging UAS Industry Trends

A number of industry trends have important implications for the cyber implications of UAS. These trends include autonomous flight capabilities, UAS traffic management, swarming, the use of AI to detect cyberattacks on UAS, increased hardware complex-

---

[5]   It is worth noting that the quality of Chinese patent data has been questioned. Schmid and Wang (2017), for example, find that China's policy of giving direct financial incentives to patent has diluted the average patent quality in China. Interviews conducted during the course of this study support this claim. Nevertheless, while there may be some, even significant, upward bias in China's patent numbers, the sheer magnitude of recent growth suggests an underlying real surge in UAS innovation in China. See Jon Schmid and Fei-Ling Wang, "Beyond National Innovation Systems: Incentives and China's Innovation Performance, *Journal of Contemporary China*, Vol. 26, No. 104, 2017.

Table 4.1
Top UAS Patenting Organizations

Organization	UAS Patents	Country of Origin	Organization Type
DJI	685	China	Firm
State Grid Corporation China	359	China	Firm (state-owned)
Boeing	345	U.S.	Firm
Ewatt Technology	193	China	Firm
Amazon	167	U.S.	Firm
Raytheon	140	U.S.	Firm
Honeywell	133	U.S.	Firm
University of Beihang	131	China	University
Guangdong Rongqi Intelligent Technology	121	China	Firm
Lockheed Martin Corporation	119	US	Firm
Avic Xian Aircraft Design and Research Institute	105	China	Firm
Haoxiang Electrical Energy	101	China	Firm
Southern Power Grid Company Limited	100	China	Firm (state-owned)
Shenzhen Autel Intelligent Aviation Tech.	98	China	Firm
IBM	96	US	Firm
University of Nanjing Aeronautics & Astronautics	91	China	University
Korea Aerospace Research Institute	85	South Korea	Government
Shenzhen AEE Aviation Technology Co. Ltd.	79	China	Firm
Northwestern Polytechnical University	79	China	University
Guangzhou Xaircraft Technology Co. Ltd.	77	China	Firm
Qualcomm	75	US	Firm
Wuhu Yuanyi Aviation Technology Co Ltd	73	China	Firm
United States Navy	70	US	Government
Zero UAV Beijing Intelligence Technology	62	China	Firm
BAE Systems	61	UK	Firm
Geer Technology Co. Ltd.	61	China	Firm
Prodrone Craft Technology Shenzhen Co.	61	China	Firm
Samsung Electro-Mechanics Co.	58	South Korea	Firm
Aerovironment, Inc.	56	Canada	Firm
China Academy of Aerospace Aerodynamics	54	China	Government

SOURCE: Clarivate Analytics, 2019.

ity and supply chains, and blockchain for UAS. Below we describe these trends and highlight some of the concerns held by industry experts related to the future of UAS and cybersecurity.

Autonomous Flight Capabilities

A typical drone requires a remote pilot to control the throttle, heading, pitch, yaw, and roll of the aircraft. The pilot may also decide when and how to control onboard equipment such as a camera. However, UAS manufacturers have developed and are continuing to develop aircraft capable of autonomous operations. Autonomous operations are defined here as those operations in which a UAS flies a trajectory that was not continuously governed by a human pilot, with an AI system having substantial planning authority and adapting to circumstances encountered during flight. This capability includes adaptively responding to unexpected situations (i.e., context-based navigation or activity-based positioning), in contrast to an automated drone that may control its subsystems according to predefined rules. As an example of a product demonstrating autonomous flight capabilities, Skydio currently markets a "self-flying camera" drone capable of following (and recording) a user while sensing and avoiding obstacles.[6] Vantage Robotics has also introduced a drone that can sense and avoid obstacles, easily comes apart based on magnetic coupling, and employs caged rotors. These capabilities have enabled it to be the only small drone to be certified by the Federal Aviation Authority (FAA) to fly over crowds. Currently, several high-profile news organizations use this drone.

Other firms have increased UAS autonomy in other ways. For example, the "return home" feature, whereby a drone autonomously returns to its operator when the feature is activated, has become ubiquitous. Similarly, GPS-enabled waypoint navigation is standard on popular UAS platforms, including the DJI Phantom, Inspire, and Mavic. Airobotics sells "automated industrial drones," advertising the fact that there is "no pilot required."[7]

Recently, researchers have made progress in miniaturizing the technologies required for autonomous UAS operations,[8] as well as in refining the technologies to allow for high-speed autonomous flight.[9] One consulting firm estimates that half of all commercial UAS flights will be autonomous by 2022.[10] The utility of autono-

---

[6]   Skydio, "The Self Flying Camera Has Arrived," webpage, 2019.

[7]   Airobotics, "Automated Industrial Drones," webpage, undated.

[8]   Jonathan Greig, "AI-Powered Autonomous Drone Could Bring New Capabilities to Agriculture, Logistics, More," *Tech Republic*, May 16, 2018; Marco Margaritoff, "World's Smallest Autonomous Drone Takes Flight in Europe," *The Drive*, May 31, 2018.

[9]   Marco Margaritoff, "MIT's NanoMap Tech Allows for Consistent, High-Speed, Autonomous Drone Navigation," *The Drive*, February 12, 2018.

[10]   Mark Huber, "Study: Half of Drone Flights to be Autonomous by 2022," *AIN Online*, March 22, 2018.

mous UAS has been noted by the U.S. military. One U.S. Air Force report notes that autonomous systems, unlike automated systems, enable a drone to be "goal-oriented in unpredictable environments and situations," and thus such systems offer clear benefits during military missions.[11]

It will likely be more difficult to detect and respond to cyberattacks involving UAS with autonomous flight capabilities. It is more likely that no human operator will be monitoring an individual drone at any given time, making it less likely that unusual or unauthorized system behavior will be noted. In addition, the planning done by autonomous systems will not necessarily be understandable or interpretable to humans. For example, research on the miniaturization of autonomy-enabling technologies are based on a "lightweight residual convolutional neural network architecture" known as DroNet.[12] These technologies process images in real time, starting with raw image data and then iteratively applying multiple convolutional filters and reducing the dimensionality of the results (e.g., by replacing data with the local maximums observed in different regions of the data). After the above steps are repeated several times, a conventional ("fully connected") neural network is typically used to process the results and to ultimately arrive at applications such as the detection of objects. It is difficult and time-consuming for a human to track such a system and to understand why, for example, a system failed to see an obstacle before crashing into that obstacle. This makes it difficult to separate anomalous and possibly malicious behavior from benign behavior. This also raises the possibility of targeted attacks that "trick" autonomous systems into behaving in unexpected ways.

UAS Traffic Management (UTM)

Air Navigation Service Providers (ANSPs) provide air traffic control and management services to pilots of conventional manned aircraft and air carriers. These organizations ensure that stakeholders adhere to established policies and procedures, ensuring safe and efficient operations. They maintain and provide others with situational awareness, at both a tactical and strategic level, of air transportation system resources and relevant information such as weather forecast data. Traffic controllers may request that pilots follow certain routes and change course as issues arise, ensuring that aircraft never get closer than a few nautical miles to one another or to obstacles. Traffic controllers manage access to shared resources such as airport taxiways and runways, avoiding inequitable outcomes.

The FAA Air Traffic Organization (ATO) acts as the ANSP in the United States and in adjacent areas above the Atlantic and Pacific Oceans. In particular, the FAA

---

[11] Robert P. Otto, "Small Unmanned Aircraft Systems (SUAS) Flight Plan: 2016-2036. Bridging the Gap Between Tactical and Strategic," Technical Report, Air Force Deputy Chief of Staff, Washington, D.C., April 30, 2016.

[12] Greig, 2018.

ATO actively monitors and manages air traffic in "controlled airspace," including airspace near relatively busy airports and at higher altitudes. For operators to use UAS in controlled airspace, coordination with ANSPs is essential. For operators to fly a drone in uncontrolled airspace where there may be another drone, helicopter, or other air traffic, some form of air traffic control and management will also be necessary. It may not be optimal, or even feasible, to employ a similar approach to manage UAS traffic, as current air traffic control and management is used to manage conventional aircraft in controlled airspace.[13] However, it is clear that some form of UAS traffic management is necessary.

Relevant systems, often labeled UTM systems, are under development now. For example, the Thales ECOsystem UTM promises "automated flight authorizations as well as real-time alerting and intervention in emergency situations."[14] The AirMap UTM platform includes "2-way communication capabilities between airspace managers and drone operators."[15] These and other systems are being developed by private firms collaborating with UAS manufacturers and with government officials. In particular, high-profile government agency efforts are seeking to "integrate" UAS and conventional aircraft traffic. These efforts often enlist private firms to help establish and then realize visions for the management of UAS traffic. The Single European Sky ATM Research (SESAR) U-space project aims to support UAS "e-registration, e-identification and geofencing" in the near-term, before later turning to UAS "flight planning, flight approval, tracking, and interfacing with conventional air traffic control."[16] In the United States, an effort led by the National Aeronautics and Space Administration (NASA) simply called UTM seeks to develop technologies that support

> airspace design, corridors, dynamic geofencing, severe weather and wind avoidance, congestion management, terrain avoidance, route planning and re-routing, separation management, sequencing and spacing, and contingency management.[17]

Note the similarity between the tasks performed by the UTM system in the NASA effort and the tasks performed by ANSPs for conventional air traffic in controlled airspace.

---

[13] Kenneth Kuhn, *Small Unmanned Aerial System Certification and Traffic Management Systems,* Santa Monica, Calif.: RAND Corporation, PE-269-RC, 2017.

[14] Thales, "Thales Launches Ecosystem UTM and Joins Forces with Unifly to Facilitate Drone Use," July 3, 2017.

[15] AirMap, *Five Critical Enablers for Safe, Efficient, and Viable UAS Traffic Management (UTM)*, white paper, January 2018.

[16] Sesar Joint Undertaking, "U-space," webpage, 2019.

[17] Joseph Rios, "Unmanned Aircraft System (UAS) Traffic Management (UTM)," National Aeronautics and Space Administration, February 15, 2019.

In NASA's vision of UTM, UAS operators send data and requests for authorization to service suppliers. These service suppliers send UAS operators notifications and information that enables the operators to improve their situational awareness. The service suppliers are not the UTM; the service suppliers communicate with the UTM, which fulfills the "Regulator/ANSP function."[18] The service suppliers also communicate with other parties, for example, to obtain aviation weather data. Figure 4.4 illustrates the communications necessary to support this vision of UTM.[19]

Note that the use of UTM systems creates new communications channels that could be exploited. These new channels provide new ways to access multiple UAS, multiple UAS operators, and ANSPs. UTM systems and service suppliers will necessarily collect data from multiple UAS and from multiple UAS operators. This collection makes UTM systems and associated networks potential targets for cyberattacks. Also note that UTM systems and service suppliers are likely to become essential to the

Figure 4.4
Communications Related to UTM

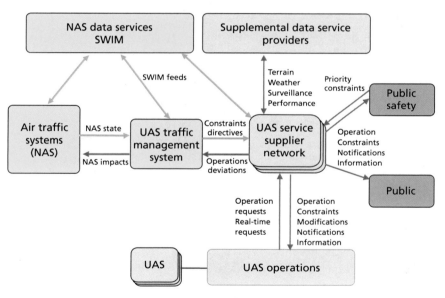

SOURCE: Kopardekar et al., 2016, Federal Aviation Administration, 2018. NAS = National Airspace System; SWIM = system-wide information management.
NOTE: Orange is regulator/ANSP function; green is UAS Operator function; blue is other stakeholders.

---

[18] Parimal Kopardekar, Joseph Rios, Thomas Prevot, Marcus Johnson, Jaewoo Jung, and John E. Robinson, III, "Unmanned Aircraft System Traffic Management (UTM) Concept of Operations," *16th AIAA Aviation Technology, Integration, and Operations Conference*, Washington, D.C.: AIAA Aviation, 2016.

[19] Kopardekar et al., 2016.

safe and efficient use of airspace. This role also highlights the need to protect them from cyberattacks.

UTM system developers are currently using the same wireless communications technologies and infrastructure being used and provided by mobile phone carriers. For example, in a recent demonstration, Qualcomm used the Long-Term Evolution (LTE) network communications standard to connect a drone with a UTM system.[20] While Qualcomm believes LTE can support some UAS use, it is promoting the wider use of 5G advanced mobile communications technologies, which it sees as necessary for "wide scale deployments of mission-critical drone use cases."[21] Qualcomm lists "strong security" as one selling point of 5G.[22]

Swarming

UAS swarming involves coordinating the operation of multiple UAS to accomplish a particularly large-scale or complex mission. Swarms are composed of multiple aircraft or (relatively few) groups of homogenous aircraft, and may be managed centrally or by a decentralized control algorithm. Generally speaking, swarms often rely on an individual small Unmanned Aerial System (sUAS) having autonomous flight capabilities and use imaging and sensors to acquire information within an environment. The swarms can act on that information to maneuver and use communications technology to receive and transmit information. Swarms also produce collective functional abilities and may exhibit unplanned solutions that would not be obvious based on the characteristics of any single drone.

The most ambitious swarm exercise to date was the U.S. Service Academies Challenge in April 2017, run by the Defense Advanced Research Projects Agency (DARPA) OFFSet (Offensive Swarm-Enabled Tactics) program, which pitted the U.S. Military Academy, U.S. Air Force Academy, and U.S. Naval Academy against one another in the skies over Camp Roberts, an Army National Guard post north of Paso Robles, California. Each academy demonstrated the offensive and defensive tactics they had developed over the course of the school year.

Two teams at a time played inside the Battle Cube, a cubic airspace 500 meters on a side, 78 meters high. Each team was given 20 fixed-wing UAS and 20 quad-rotor UAS and, under the rules of play, could field a mixed fleet of up to 25 UAS for each of two 30-minute battle rounds. Each team had to protect its "flag" (a large, inflatable ground target) while trying to score the most points before time ran out. The benefits of swarming, as shown in this exercise, include improved performance on tasks that

---

[20] Maged Zaki, "Path to 5G: Building a Highway in the Sky for Autonomous Drones," Qualcomm Technologies, November 9, 2016.

[21] "Leading the World to 5G: Evolving Cellular Technologies for Safer Drone Operation," Qualcomm, September 6, 2016.

[22] "Leading the World to 5G: Evolving Cellular Technologies for Safer Drone Operation," 2016.

can run in parallel, the ability to perform multiple actions simultaneously in different locations, and increased fault tolerance.[23] Drawbacks include the potential for individual aircraft in a swarm to interfere with one another, uncertainty regarding what other aircraft are doing, and the overall cost of deploying and managing the swarm.[24] UAS swarms are also often composed of aircraft that have limited capabilities on their own; for example, the capability to recognize or counteract malicious instructions. Communications pose a particular challenge and will be discussed later in this section of this report.

DARPA is currently running another program, named Gremlins, that is investigating the launch of sUAS swarms from conventional aircraft,[25] as well as the above-mentioned OFFSET program, whose ultimate goal is equipping infantry forces with 250 or more sUAS.[26] The Russian military has claimed that its base in Syria was attacked by a sUAS swarm.[27] A U.S. Air Force technical report notes that swarming offers advantages, for example "the ability to triangulate targets when seen from three or more vantage points."[28]

An extensive academic literature now exists on swarming, particularly on the control architectures and algorithms required for swarming. For example, one article proposes "a decentralized approach based upon information-theory and distributed data fusion which enable the scale up to large numbers of collaborating Small Unmanned Aerial Systems (sUAS) platforms."[29] Other research efforts draw inspiration from flocks of birds and schools of fish.[30] The authors note that geolocating a target is easier and faster when considering sensor readings from multiple sUAS.

Swarming requires autonomous flight capabilities, raising concerns related to cybersecurity that were mentioned earlier in this chapter. Swarming also requires some form of management over the swarm. A centralized management authority would be a reasonable target for a cyberattack because access to or control over the authority would grant an attacker with data on or control over hundreds of aircraft at once. Even decentralized management schemes rely on extensive communications that could be attacked to help a malicious actor understand or alter operations. For example, note

---

[23] Ronald Arkin, *Behavior-Based Robotics*, Cambridge, Mass.: MIT Press, 1998.

[24] Arkin, 1998.

[25] Scott Wierzbanowski, "Gremlins," Defense Advanced Research Projects Agency, undated.

[26] "OFFensive Swarm-Enabled Tactics (OFFSET)," Defense Advanced Research Projects Agency, undated.

[27] David Axe, "How Russia Says It Swatted Down a Drone Swarm in Syria," *Motherboard*, January 12, 2018.

[28] Otto, 2016.

[29] Raj P. Malhotra, Michael J. Pribilski, Patrick A. Toole, and Craig Agate, "Decentralized Asset Management for Collaborative Sensing," *Proceedings Volume 10194, Micro- and Nanotechnology Sensors, Systems, and Applications IX*, Anaheim, Calif.: SPIE Defense + Security, May 18, 2017.

[30] Prabir Barooah, Gaemus E. Collins, and João P. Hespanha, "GeoTrack: Bio-Inspired Global Video Tracking by Networks of Unmanned Aircraft Systems," *Proceedings of the SPIE*, Vol. 7321, May 2009.

the many links between the authors' proposed "Decentralized Asset Manager" and other assets in Figure 4.5.[31] As Higgins, Tomlinson, and Martin (2009) note, "Swarm robots can interact either explicitly, or implicitly," but, either way, "any open implicit or explicit communication method can be jammed, intercepted or otherwise disturbed relatively easily by an attacker."[32]

Technologies that support the communications necessary for swarming include radio frequency (RF) and LTE technologies. There are known cyber issues associated with these technologies. For example, LTE uses "commodity hardware and software"

Figure 4.5
Decentralized Asset Manager for Swarming

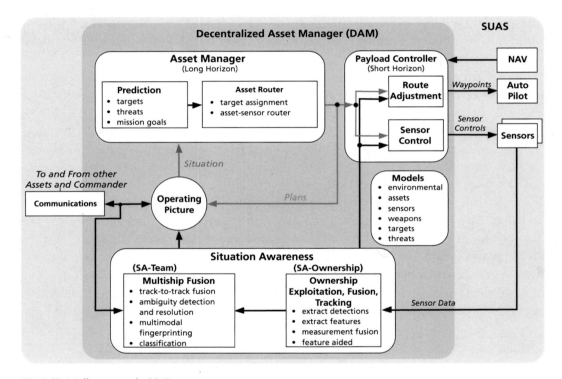

SOURCE: Malhotra et al., 2017.

---

[31]   Malhotra et al., 2017.

[32]   Fiona Higgins, Allan Tomlinson, and Keith M. Martin, "Survey on Security Challenges for Swarm Robotics," *2009 5th International Conference on Autonomic and Autonomous Systems*, Valencia, Spain: ICAS, April 2009.

that have known vulnerabilities.[33] Jamming remains an "unaddressed threat" capable of preventing successful transmission of RF and LTE-enabled signals.[34] Software- (SW-) defined radio would help mitigate vulnerabilities by facilitating dynamic sensing and adaption to the RF environment. Multiaperture techniques provide further options to leverage spatial diversity and coherence to steer transmissions toward the intended receivers and make null sources of interference.[35] To improve RF propagation, distributed coherence and multi antenna techniques can be used to maintain awareness of the RF environment.[36]

Examining the corpus of journal articles on UAS swarming en masse reveals two additional insights.[37] First, scientific research into UAS swarming has accelerated in recent years and appears to be on an exponential growth trajectory. Second, as was observed in the case of overall UAS patenting, the scientific literature on UAS swarming is increasingly dominated by Chinese organizations.

Figure 4.6 depicts the annual world output of scientific research in the field of UAS swarming.[38] Since the first journal article on the subject was published in 2002, there has been exponential growth in this subfield. In fact, fitting an exponential function to the data reveals a relatively good fit (r-squared of 0.88). Using the exponential function to forecast future scientific output in this field suggests estimated publication counts in 2018, 2019, and 2020 of 164, 214, and 279, respectively.

The second insight is that Chinese organizations are responsible for the majority of research output. Table 4.2 shows the author affiliations of the most prolific authors in the subfield. Figure 4.7 depicts the collaborative (coauthorship) network of author affiliations. The network graph illustrates that Chinese researchers are establishing strong communities of scientific practice. The presence of knowledge networks is indicative of a healthy national scientific community for a given sector; they indicate the presence of channels of interorganization information flow. Of the eight communities of coauthorship observed in the network, four are dominated, in terms of authorship slots contribution, by Chinese organizations. The collaborative relationship between Beihang University (previously Beijing University of Aeronautics and Astro-

---

[33] Jeffrey Cichonski and Joshua Franklin, "LTE Security – How Good Is It?" presentation given at the 2015 RSA Conference, San Francisco, Calif., 2015.

[34] Cichonski and Franklin, 2015.

[35] Harry L. Van Trees, Kristine L. Bell, and Zhi Tian, *Detection, Estimation, and Modulation Theory: Part I– Detection, Estimation, and Filtering Theory*, 2nd ed., Hoboken, N.J.: Wiley, 2013.

[36] Van Trees, Bell, and Tian, 2013.

[37] Currently, the scientific and technological progress in the field of UAS swarming is located primarily in the academic literature. Adding swarm* to our primary patent search yielded 43 results (43/19,333). Adding swarm* to the academic literature search produced 564 results (564/15,843).

[38] Result based on Web of Science search. The swarming subset is constructed by simply adding "swarm*" to the original query of "TS= (UAV AND unmanned) OR TS= (UAS AND unmanned) OR TS = ("unmanned aerial") OR TS= (drone AND unmanned.)"

Figure 4.6
Number of Publications on UAS Swarming

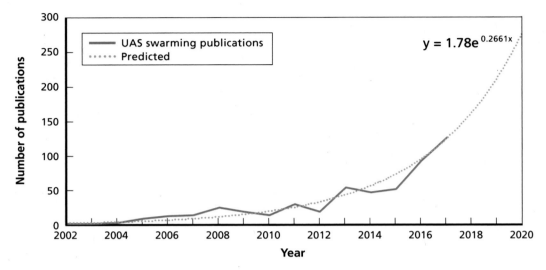

SOURCE: Clarivate Analytics, undated.

Figure 4.7
Network of Swarming Publications

NOTES: IOSB = Institute of Optronics, System Technologies and Image Exploitation; Univ = university; Natl = national; Inst = institute.

Table 4.2
Affiliation of Authors of Swarming Publications

Organization	UAS Swarming Publications	Country of Origin	Organization Type
Beihang University	51	China	University
Beijing Institute of Technology	19	China	University
Technical University of Dortmund	12	Germany	University
Cranfield University	11	UK	University
Tsinghua University	11	China	University
United States Air Force	10	U.S.	Government
De La Salle University	9	Philippines	University
Nanyang Technological University	8	China	University
Harbin Institute of Technology	7	China	University
PLA National University of Defense Technology	7	China	University

nautics) and Tsinghua University is, by a significant margin, the strongest in the entire coauthorship network.

Using AI to Detect UAS Cyberattacks

Some emerging technology trends could potentially help mitigate the risks outlined earlier in this chapter and other sections of the report associated with intrusion detection and tampering with system functionality. One such trend is the use of ML and AI techniques and tools to detect cyberattacks in real time.

The intrusion detection problem in cybersecurity involves identifying malicious use and policy violations among those using a cyber system. Numerous researchers have proposed using ML and AI methods on network and system communications. Buczak and Guven, in a 2016 survey article on the topic,[39] note that techniques that have been applied include neural networks, decision trees, ensemble learning, and support-vector machines, among others. These are all common ML methods.[40] These techniques are able to identify various types of intrusions and attacks by monitoring network traffic. The authors find that the "biggest gap" preventing more-successful applications of such techniques, or even a fair comparison among them, is the lack of training data—datasets that contain known intrusions and could be used to train algo-

---

[39] Anna L. Buczak and Erhan Guven, "A Survey of Data Mining and Machine Learning Methods for Cyber Security Intrusion Detection," *IEEE Communications Surveys & Tutorials*, Vol. 18, No. 2, Second Quarter, 2016.

[40] Trevor Hastie, Robert Tibshirani, and Jerome Friedman, *The Elements of Statistical Learning*, New York: Springer, 2001.

rithms to detect future threats.[41] Jones and Straub note that several "machine learning techniques have been used to approach the intrusion detection problem."[42] The authors go on to introduce a methodology specifically for intrusion detection for "autonomous robots" that includes one component that identifies "signatures in robot decisionmaking," essentially patterns in activity that reveal something important about how the robot is behaving, and another based on a "deep neural network that is trained to detect commands that deviate from expected behavior."[43]

Signature identification is based on a database of previously identified signals. The overall methodology involves a combination of a rule-based system (for identifying "signatures") and a neural network (for determining whether there has been an intrusion); similar hybrid approaches have been proposed relatively frequently in this domain. Such an approach could be promising for detecting cyber-UAS attacks, though, again, the lack of training data presents a challenge.

A small but growing body of literature looks more specifically at intrusion detection for UAS. The topic is often linked to fault detection, the related idea of identifying, often in real-time, when something has gone wrong with a component of a UAS. For example, Abbaspour et al. (2017) propose an approach based on a neural network that detects both faults and false data injection attacks on "unmanned quadrotor sensors."[44] Gil-Casals, Owezarski, and Descargues (2013) describe a support-vector machine-based method for "autonomous detection of cyberattacks on airborne networks."[45] Airborne networks here include aerial vehicles that communicate to one another, or to ground-based systems. In her longer doctoral thesis,[46] Gil-Casals has defined a risk-assessment framework for cyber threats to such networks. One part of Gil-Casals' thesis lists features that can be used to describe network traffic quantitatively and research articles where the features were proposed. Sedjelmaci and Senouci (2018) describe a "security framework" for mitigating cyberattacks on airborne networks.[47] The authors describe how different data are relevant for different purposes; for

---

[41] Hastie, Tibshirani, and Friedman, 2001.

[42] Andrew Jones and Jeremy Straub, "Using Deep Learning to Detect Network Intrusions and Malware in Autonomous Robots," *Proceedings Vol. 10185, SPIE Defense + Security, Cyber Sensing,* April 2017.

[43] Jones and Straub, 2017.

[44] Alireza Abbaspour, Michael Sanchez, Armen Sargolzaei, Kang Yen, and Nalat Sornkhampan, "Adaptive Neural Network Based Fault Detection Design for Unmanned Quadrotor Under Faults and Cyber Attacks," *2017 25th International Conference on Systems Engineering,* Las Vegas, Nev.: IEEE, August 2017.

[45] Silvia Gil-Casals, Philippe Owezarski, and Gilles Descargues, "Generic and Autonomous System for Airborne Networks Cyber-Threat Detection," *2013 IEEE/AIAA 32nd Digital Avionics Systems Conference (DASC),* October 2013, pp. 4A4.

[46] Silvia Gil-Casals, *Risk Assessment and Intrusion Detection for Airborne Networks,* dissertation, Toulouse, France: INSA Toulouse, 2014.

[47] Hichem Sedjelmaci and Sidi Mohamed Senouci, "Cyber Security Methods for Aerial Vehicle Networks: Taxonomy, Challenges and Solution," *Journal of Supercomputing,* Vol. 74, No. 10, October 2018.

example, how "data injection rate" can be used to detect "wormhole and black hole" attacks while "signal strength intensity" can be used to detect "jamming" and "GPS spoofing" attacks. Aktaş, Gemci, and Yağdereli (2015) similarly classify threats and then propose general guidance for responding to such threats.[48] The authors note that UAS "must be equipped with defensive capabilities and measures such that they can be able to respond automatically and dynamically to both accidental and deliberate defects and attacks."[49] Another survey article, by Shakhatreh et al. (2018), organizes 16 relevant previously published research articles based on the "attack vector" and the "proposed countermeasure" that each article considers and introduces.[50] One point that the authors bring up is the potential for "adversarial attacks on the employed machine learning techniques."[51] It makes sense that if a particular ML or AI technique or tool is widely used to detect intrusions: then attackers may develop strategies for deceiving the technique or tool itself.

Hardware and Fabrication Complexity

Commercially available UAS are becoming increasingly sophisticated. One consequence of increased hardware and software complexity is that it becomes harder to track and verify the various processes that are operating on a particular platform. Additionally, the introduction of automation and computer-aided design into the fabrication of UAS and components means that oversight of and responsibility for system components may be increasingly difficult to trace. Specific components within communication (e.g., Bluetooth, GPS, WiFi, USB), sensor (e.g., infrared, radar, lidar), and flight control (e.g., AI for autopilot, dynamic following, swarm behavior) systems may be vulnerable, and exposure to these kinds of vulnerabilities may be increasingly difficult to uncover.

In essence, the drone acts as a network of components that communicate with each other and have various levels of interdependencies. Like software viruses, hardware components, mostly integrated circuits, can be injected with malicious code to cause unwanted behavior based on defined input sequences. This is already occurring in UAS.[52] Most public-use UAS models consist of third-party intellectual property blocks, one or more of which may be contaminated during design, fabrication, configuration, or even the use phase. New technologies provide new attack vectors for

---

[48] Ziya A. Aktaş, Cemal Gemci, and Eray Yağdereli, "A Study on Cyber-Security of Autonomous and Unmanned Vehicles," *Journal of Defense Modeling and Simulation*, Vol. 12, No. 4, October 2015.

[49] Aktaş, Gemci, and Yağdereli, 2015.

[50] Hazim Shakhatreh, Ahmad Sawalmeh, Ala Al-Fuqaha, Zuochao Dou, Eyad Almaita, Issa Khalil, Noor Shamsiah Othman, Abdallah Khreishah, and Mohsen Guizani, "Unmanned Aerial Vehicles: A Survey on Civil Applications and Key Research Challenges," arXiv.org, arXiv:1805.00881, 2018.

[51] Shakhatreh et al., 2018.

[52] Swati Khandelwal, "MalDrone — First Ever Backdoor Malware for Drones," *Hacker News*, January 27, 2015.

infecting these hardware systems. For example, the first complete sabotage attack on a three-dimensional (3D) printed drone propeller was demonstrated by researchers at Ben-Gurion University in Israel using malicious manipulation of blueprints in 2016.[53] Computer-aided design (CAD) tools used to manufacture most electronic circuits today provide similar entry points for attack, enabling, for example, memory corruption through a malicious drawing file.[54]

For hardware threat identification, the cascading nature of failures of these components adds complexity and may be undiscoverable except through the use of specialized simulation software. Malicious logic can lead to such unwanted scenarios as causing the system to output data to the wrong port or address (information leakage), monitoring and modifying the system's output data (tampering), or disabling the system by changing the system's internal timing or control. All these can be done by changing or adding internal logic in such a way that it is unlikely to be detected by traditional testing and verification tools.

Blockchain for UAS

Blockchain technology also has relevance for the cybersecurity implications of UAS proliferation. A blockchain is a list of encrypted records stored in a distributed public ledger. The first high-profile blockchain records transactions involved the Bitcoin digital currency. Each user of Bitcoin maintains a copy of the blockchain, and all copies of the blockchain are updated regularly. The company Blockchain develops tools for managing Bitcoin.[55] It is worth noting that Bitcoin has been labeled a vehicle for money laundering for organized crime and terrorist organizations,[56] and it may be used to fund attacks of the type considered in this report. Other blockchains have been developed and are used for other digital currencies and, more generally, for recording other types of events. Blockchain has proven particularly adept at supporting automated, secure registries of transactions.

Many firms have proposed using a blockchain to record UAS deliveries of goods (e.g., Dorado,[57] Walmart).[58] Others have proposed using blockchain to manage communications between UAS operators and "aviation authorities," including those who might manage or utilize UTM systems (e.g., applied Blockchain).[59] As one promo-

---

[53] Sofia Belikovetsky, Mark Yampolskiy, Jinghui Toh, Jacob Gatlin, and Yuval Elovici, "dr0wned–Cyber–Physical Attack with Additive Manufacturing," *WOOT '17 Proceedings of the 11th USENIX Conference on Offensive Technologies*, Vancouver, Canada, August 2017.

[54] Votiro Labs, "AutoCad Security Bug Still a Risk, It Turns Out," January 3, 2018.

[55] Blockchain, "About," webpage, 2019.

[56] Antonia Ward, "Bitcoin and the Dark Web: The New Terrorist Threat?" The RAND Blog, January 22, 2018.

[57] Dorado, "Dorado ICO," webpage, undated.

[58] Robert Hackett, "Wal-Mart Explores Blockchain for Delivery Drones," *Fortune*, May 30, 2017.

[59] Applied Block Chain, "Applied Blockchain: Building Future-Proof Blockchain Applications," webpage, 2018.

tional website notes, many UTM research and development efforts seek "to create a system that does not require constant human monitoring and surveillance and can still ensure the authenticity, safety, security and control of the drones in the low-altitude airspace ... Blockchain can facilitate all these features."[60] Academic researchers have begun to study the potential use of blockchain in UTM. One recent article proposes a UAS "Traffic Information Exchange Network" based on blockchain.[61]

The decentralized, duplicative nature of blockchain and, in particular, by-design redundant checks on data quality used in normal functioning of the block chain make it difficult for an attacker to adjust or delete records. This feature could make it difficult for an attacker to, for example, modify flight plans within a UTM system. The use of encryption supports secure communications. This encryption could make it more difficult for an adversary to gather information from a drone. On the other hand, the distributed nature of data storage raises security concerns. Blockchain technology could be used to record, gather, and easily search for information on where a drone has been or what it has done. This aspect could enable an attack or applications that identify rogue UAS or unexpected patterns in UAS use.

An article in the *Harvard Business Review* claims that blockchain is "decades from reaching its full potential" due to the relative novelty and complexity of the technology.[62] A study published in October 2017 that included a survey of supply chain and logistics professionals found that just 20 percent had implemented any form of "blockchain solutions."[63] The primary barriers to the wider use of blockchain, according to the professionals surveyed, included: "regulatory uncertainty," the need for different parties to agree to and utilize a common system, and a "lack of technological maturity."[64] Data security was also a major concern. It is currently impossible to tell what effect blockchain will have on the cybersecurity of commercially available UAS, but it is definitely a technology and trend that warrants further scrutiny in the next several decades.

---

[60] Amit Ganjoo, "Blockchain and Drones—The Reality," ANRA Technologies, February 13, 2018.

[61] Hsun Chao, Apoorv Maheshwari, Varun Sudarsanan, Shashank Tamaskar, and Daniel DeLaurentis, "UAV Traffic Information Exchange Network," *2018 Aviation Technology, Integration, and Operations Conference*, Atlanta, Ga.: AIAA Aviation Forum, 2018, p. 3347.

[62] Marco Iansiti and Karim R. Lakhani, "The Truth About Blockchain," *Harvard Business Review*, January–February 2017.

[63] Niels Hackius and Moritz Petersen, "Blockchain in Logistics and Supply Chain: Trick or Treat?" *Proceedings of the Hamburg International Conference of Logistics (HICL)*, Vol. 23, Hamburg, Germany: Hamburg International Conference of Logistics (HICL), 2017.

[64] Hackius and Petersen, 2017.

Improved Hacking and Malware Delivery Support

In addition to some of the trends described above, which may help improve drone performance, security, or the ability to identify and respond to attacks, new technologies are also supporting attackers. Many models of drones can be easily disabled or even co-opted by a malicious actor, and readily available technologies help support such actions.

For example, SkyJack is an application that can exploit the weakly encrypted WiFi access ports of civilian UAS.[65] It enables a drone or computer to seek out and take control of other drones within proximity. SkyJack targets Parrot AR.Drone 2.0 drones through network security software. Aircrack-ng, a security software, is used to identify a target's wireless networks and clients through a WiFi access point. After obtaining access to a network and its clients, the application proceeds to take control of the target by disconnecting all the connected clients and requesting control of the now "uncontrolled" drone. What is notable about SkyJack is that it is scalable. A single instance of SkyJack can compromise multiple targets, and it is likely adaptable to multiple different types of drones because its source code is public and exploits a common underprotected WiFi access port on drones.

Maldrone is a malware that can be used remotely or locally to gain access to a drone's operating system without the owner's knowledge. It was designed to be "the first ever drone backdoor" and works by inserting itself as a man-in-the-middle in the communication between the drone's automation software and hardware.[66] It listens to traffic from real ports and sends a combination of falsified and legitimate commands to the command program using proxy ports. The source code for Maldrone is open to the public and could be easily adapted to infect many types of drones. Maldrone can also be uploaded into drones remotely, potentially enabling malicious actors to capture multiple drones at once. Thus, a malicious actor can use Maldrone to send various commands to an infected drone ranging from remote surveillance to gaining control over UAS flight.

Agent-Based Modeling and Simulation for Drone Defense

This report does not cover the large literature on counter-UAS. However, in some cases, physical defense against drones could in fact be useful in preventing cyberattacks where drones are used to deliver malware or gain access to a system through proximity. Therefore, the development of improved simulation technology to aid in building physical drone defense systems is notable. Technologies such as Map-Aware Non-Uniform Automata (MANA), developed by the New Zealand Defense Technology Agency, are increasingly being used to understand threats and the efficacy of various defense approaches. Such simulations allow users to study the effects of drone

---

[65] Samy Kamkar, "SkyJack," webpage, December 2, 2013.

[66] Charlie Osborne, "Maldrone: Malware Which Hijacks Your Personal Drone," *ZDNet.com*, January 27, 2015.

characteristics (e.g., firing rate, flight speed, payload) and defense characteristics (e.g., communications systems, identification accuracy, latency) on attack outcomes, providing decisionmakers with insights into their ability to counter drones, both today and in light of possible future technological advances for either the attacker or defender.

Table 4.3 provides a summary of some of the key UAS-related trends discussed in this section. This table also identifies the STRIDE dimensions which may come under increased risk due to these technological developments. The table also specifies the attack vector which may be opened due to these technological trends.

Industry Trends: Conclusion

Experts in cybersecurity and UAS share many of the concerns regarding the trends highlighted above. While working toward market introduction, many firms produc-

Table 4.3
Summary of Key UAS Features and Trends

Trend	Key UAS Feature	STRIDE Taxonomy Threat	Vulnerabilities and Attack Vectors
Simplified Control and Operation	Camera view-based flight; following target on camera	Repudiation and Information Disclosure	Third-party monitoring of user activities
	Gesture and speech-directed flight control	Elevation of Privilege and Tampering	Alteration of factory-installed configurations
Self-Operation and Vigilance	Location or sensor-based payload manipulation (e.g., crop spraying, medical supply delivery)	Elevation of Privilege	Intercept of payload usage or delivery
	Swarm drone maneuvers; multi-UAS operations	Elevation of Privilege and Tampering	Scaled-propagation of operational errors
	Preplanned hovering; patrol routines	Spoofing or Tampering	Override of authentic GPS signal or uploaded navigation files
Self-Maintenance and Protection	High-speed obstacle avoidance	Spoofing and Denial of Service	Sensor saturation or interference for obstruction of "view"
	Auto-docking; recharge; return to home	Repudiation and Information Disclosure or Spoofing and denial of service	Third-party monitoring of user activities and sensor interference for failure to register "home" state

ing public-use UAS have neglected to meaningfully address hardening their platforms to resist a cyberattack. However, as these drones begin to be adopted for DHS missions,they will have to comply with emerging standards around cybersecurity and data protection. Many in the industry will be waiting for standards to be released so that they can differentiate themselves in the market and capture market share in the USG and local law enforcement markets.

One possible next step for helping to address some of the concerns of both industry leaders and policymakers may be the establishment of more-standardized UAS test methodology and approach. It is clear that other federally funded research and development centers (FFRDCs) and UAS experts would support such a strategy.[67] To that end, we recommend a unified and partnered approach across the government to establish a UAS test range that can test and verify vendor claims to actual public-use UAS performance.

This chapter has described several technologies and trends that may have important implications for cybersecurity and, in particular, the cybersecurity of UAS. Analysts expect UAS use to grow rapidly and UAS operations to become more sophisticated. For example, autonomous operations–planning driven by an optimization goal and swarming will become more commonplace. Managing large amounts of UAS traffic will require increased communications bandwidth and the use of specialized UTM systems. These developments will produce opportunities for cyberattacks involving UAS. However, at the same time, new applications of ML and AI may produce more effective intrusion-detection systems. Blockchain technology may become widespread in the UAS world, ensuring certain communications are encrypted, but also increasing the amount of data available to would-be attackers. It will be important for DHS, in particular, to track developments in the use of UAS technology.

---

[67] Sandia Technical Staff, interviews, 2010–2014.

CHAPTER FIVE

UAS, Cybersecurity, and the Department of Homeland Security

This chapter looks at the ways in which UAS cyber vulnerabilities and capabilities can affect DHS, as well as potential and current DHS efforts to cope with UAS cyber concerns. We first examine attacks against DHS with UAS either as the target or the vector. Then we examine the reverse: potential offensive use of UAS by DHS. Finally, we look at how DHS can address UAS cyber concerns by examining how DHS components and offices can mitigate these concerns, what projects are currently underway within DHS related to UAS cybersecurity, and what current and future policies may help or hinder DHS decisionmakers.

Attacks Against DHS Assets

DHS-Operated UAS as Target

As discussed in Chapter Three, splitting UAS cyber vulnerabilities into two primary categories— those that target the UAS itself, and those that use UAS as a cyberattack vector—can confer an analytic advantage. While a small subset of DHS offices and components are currently vulnerable to one or both types of attacks, the increased ubiquity of public-use UAS and associated ease of access to these systems will further raise the level of risk to DHS, across multiple fronts.

Four DHS components have documented historical use of UAS in their day-to-day activities: the U.S Coast Guard (USCG),[1] CBP,[2] FEMA,[3] and the Cybersecurity and Infrastructure Security Agency (CISA).[4] Most of these organizations also plan to expand their use of UAS: USCG plans to outfit its full fleet of National Security Cut-

---

[1] U.S. Coast Guard, "Unmanned Aircraft System," undated.

[2] U.S. Department of Homeland Security, Customs and Border Protection, 2018.

[3] "Drone Use Reaches 'Landmark Level' in Harvey Disaster Response," *InfoGram*, Vol. 17, No. 37, September 14, 2017.

[4] DHS, 2017.

ters with small UAS,[5] CBP is exploring the use of small UAS to complement its Predators and other assets,[6] and FEMA has all also expressed an interest in expanding its use of UAS.[7] In addition, ICE is in the initial stages of understanding how UAS can enhance its agents' abilities in the field.[8]

DHS components are using and will continue to use a mix of both DoD-developed (Predator and ScanEagle) and commercially developed UAS. However, with the exception of USCG, all components plan to invest in commercially available UAS going forward. This means CBP, FEMA, CISA, and ICE assets will all be vulnerable to the types of attacks described in Chapter Three, in the following ways:

- CBP may lose intelligence, surveillance and reconnaissance (ISR) capabilities, creating visual blind spots for smuggling or other nefarious actions at borders and ports. CBP may also chose to employ UAS platforms for other activities in the future: For example, compromised UAS systems could impact chemical, biological, radiological, nuclear, and explosives (CBRNE) scanning at ports, where compromised UAS systems could prevent CBP agents from completing their duties, cause significant financial damage by delaying cargo movement while the system is fixed, or even send false "safe" readings of dangerous cargo. Compromised UAS could also create unknowable risk if the CBP operator is unaware of the breach.
- Compromised FEMA UAS may reduce capability to identify, reach, or supply individuals in peril or medical distress in disaster zones. This may happen both because the compromised UAS asset is no longer capable of performing as intended, and because the drone could be used to degrade other aerial operations such as helicopter flights or activity of other UAS. Compromised FEMA UAS may also lead to degraded situational awareness if UAS are used for ISR in disaster zones.
- Compromised CISA UAS would degrade the ability of the agency to conduct critical infrastructure inspections in some cases, and could be used in a cyber-physical attack to damage the critical infrastructure it was meant to survey. Compromised UAS could also create unkowable risk if the CISA operator is unaware of the breach.
- Finally, ICE intends to use UAS to reduce risk during raids. Compromised ICE UAS will reduce overall capability, require fallback to less-familiar concepts of operation (CONOPS), and increase the level of risk to the agents in the field.

---

[5]   USCG, undated.

[6]   U.S. Department of Homeland Security, Customs and Border Protection, "CBP to Test the Operational Use of Small Unmanned Aircraft Systems in 3 U.S. Border Patrol Sectors," September 14, 2017.

[7]   U.S. Federal Emergency Management Agency, "Transforming Cardiac Emergency Care with Drone Delivery of AEDs," webpage, 2018.

[8]   Betsy Woodruff, "ICE Wants Drones," *Daily Beast*, April 27, 2018.

Compromised UAS may even create unknowable risk if the ICE operator is unaware of the compromise.

DHS Attacked with UAS as Vector

Nearly all DHS components and offices could become victims of a drone-led botnet or data exfiltration attack. They all have physical locations where sensitive data and wireless networks are prevalent, making them targets for these types of attacks. UAS that have loitering capabilities—for example, those that can land and take-off again after some period of time—allow this type of covert attack, further increasing risk to unhardened systems.

As the ubiquity of connected devices grows, the danger of a drone-injected worm or similar attack, as discussed in Chapter Three, also increases. This attack vector need not be limited to DHS networks and connected devices, because DHS employees' personal devices or home networks could also be access points for nefarious code to gain entry to DHS systems either wirelessly or by an employee connecting an infected device to a DHS laptop.

DHS Offensive Cyber Actions with UAS as Vector

While DHS may be targeted by adversary UAS as discussed above, it can, of course, use UAS to perform its own offensive cyber actions. DHS may use UAS to observe targets or suspected adversaries covertly; for example, drones may be able to gain access to a local network suspected of coordinating smuggling operations. This type of access could be used in overt offensive ways as well, such as disabling or feeding false information to networked security cameras and alarms immediately before a raid by ICE.

UAS-on-UAS offensive cyber operations could be useful as well. CBP could employ UAS to catch or disable UAS involved in smuggling when ground or manned platforms lack the maneuverability or quick response required to pursue these fleeting targets. FEMA could choose to employ a UAS to jam or disable drones in a disaster zone to allow clear passage for helicopters. A more-selective version could involve a FEMA drone attacking only UAS that are not emitting an identifying friend-or-foe code. Finally, DHS could employ its own "guard dog" UAS unit to patrol sensitive physical locations and disable adversary UAS found in the area in proximity to any DHS component at risk of infiltration or data extraction by adversary UAS. This attribute may hold special appeal to CISA, around critical infrastructure, and the U.S. Secret Service, around national special security events. However, as discussed in the previous section, nearly all DHS offices and components are potential targets due to wireless networks and future IoT concerns.

DHS Components and Offices as Mitigators

Certain components and offices within DHS are well-positioned to be leaders in mitigating these risks. In particular, the CISA and Science and Technology Directorate can develop technical solutions, while the Management Directorate; Office of Operations Coordination; and Office of Strategy, Policy, and Plans can approach the problem from a policy perspective.

Mitigation strategies may come from three approaches. For UAS as target, three recommendations may help DHS better position itself in light of current and future trends: (1) DHS should engage with the UAS manufacturing industry and (2) DHS purchasing authorities and policymakers can create an environment in which DHS operates secure UAS with minimal possible risk. For UAS as vector (3), all DHS components and offices can reduce risk by securing and hardening their networks and developing defensive CONOPS to randomize UAS deployment configurations, utilize secondary data sources for network tampering detection, and operationalize functional integrity checks within mission timelines in the field.

As noted in Chapter Three, UAS as targets have multiple vulnerabilities, and these abound across potential attack vectors: poor passphrase security, known default settings, and unprotected ad-hoc networks have all been entry routes for attackers. Vulnerabilities have also been shown throughout various subsystems and their communication links. DHS should engage with industry leaders through the Office of Partnership and Engagement and the Science and Technology Directorate (DHS S&T) to develop standards for secure public-use UAS. While such standards will not be adopted industrywide, especially by foreign manufacturers, this process can encourage manufacturers to create UAS that are more responsive to security concerns. It should be noted, though, that adversaries will then have the same ability to purchase these cyber-hardened UAS. This measure and countermeasure cycle, typical of offense and defense competition, should encourage DHS S&T to invest in UAS cyber vulnerability threats and mitigation research, where permitted,[9] to ensure that DHS remains at the forefront of knowledge on these topics. This could also include sponsored research or prizes for the development of specific capabilities to draw in academic and commercial participation.

DHS policymakers and purchasing authorities must follow up research and industry engagement by ensuring that DHS purchases UAS that satisfy security concerns. The Office of Strategy, Policy, and Plans and the Management Directorate both have influence on policy department-wide. All components and offices that will use UAS (CBP, FEMA, CISA, ICE, and potentially USCG, as discussed earlier) must also support UAS security policy and create incentives for acquisition to purchase secure UAS. Federal law enforcement training centers and the Office of Partnership and Engagement can also be called upon to educate and influence law enforcement agencies to

---

[9]   DHS is restricted on using and testing certain types of mitigations unless policy changes are implemented by Congress. See Kirstjen M. Nielsen, "The U.S. Isn't Prepared for the Growing Threat of Drones," *Washington Post*, July 4, 2018.

adopt similar policies when making their own UAS purchases, thereby lessening the risk that these assets could be commandeered and turned against DHS assets.

To reduce UAS as vectors, all DHS offices and components should ensure their networks are secure and robust to fend off UAS-launched cyberattacks. As with UAS as targets, the Office of Strategy, Policy, and Plans and the Management Directorate can create incentives and policy to ensure this happens. DHS should also educate employees on the dangers posed by IoT and unsecured devices to reduce the risk of nefarious actors gaining entry from compromised information technology platforms. When assessing UAS countermeasures, including the potential use of counter-UAS, UAS components should engage with the Office of General Counsel and the Privacy Office to ensure these countermeasures and their use are legal.[10]

Relevant DHS Projects

While we have suggested how DHS may choose to act to protect itself from UAS cyberattack issues in the future, it is important to note that DHS is already addressing the problem. DHS has recently run (and is currently running) several projects that are relevant to this report. DHS has a Program Executive Office for Unmanned Aerial Systems (PEO UAS) in its Science and Technology Directorate.[11] One function of this office involves "enabling" UAS for integration and employment in DHS activities. A recent fact sheet put out by the PEO UAS on this topic notes that a recently completed study and simulation of sUAS employed in an environment where GPS technologies were threatened revealed weaknesses.[12] Under the same "Resilient GPS and Communication" heading, the PEO UAS has planned to demonstrate "modular communication system for sUAS in a denied GPS environment" in fiscal year 2018. PEO UAS also reports that it is working with NASA to test and evaluate UTM systems.

As part of its 2016 and 2017 First Responder Electronic Jamming Exercises, DHS sought to "[d]emonstrate [and] [a]nalyze the impacts of [e]lectronic threats on sUAS technology."[13] DHS is concerned about the growing availability of cheap devices that can disrupt electronic communications systems, and sUAS technologies are potentially attractive targets. The exercises provided DHS with an opportunity to study vulnerabilities in a realistic environment where threats were present.

---

[10] Third-party UAS (e.g., those owned by hobbyist civilians) that are stolen or taken over by adversaries could potentially contain personally identifiable information. The Privacy Office should advise on countermeasures that could potentially be sued against this type of UAS.

[11] U.S. Department of Homeland Security, "Program Executive Office for Unmanned Aerial Systems," webpage, undated.

[12] United States Department of Homeland Security, "Enabling Unmanned Aircraft Systems," DHS Science and Technology Directorate, May 3, 2017.

[13] Tim Bennett, "Air Based Technologies," presentation at the ATCA Aviation Cybersecurity Conference, 2017.

Current and Future Policy Related to DHS and UAS Cybersecurity Concerns

As recently highlighted by then–DHS Secretary Kirstjen Nielsen,[14] current policy restricts the efforts DHS may take to uphold its responsibilities to defend the homeland and its own interests against UAS attacks, including cyberattacks. However, recently proposed legislation indicates that in the near future, these restrictions may be loosened, enabling DHS to more freely address UAS-related cyber threats and launch offensive cyber options from UAS platforms, although certain areas of concern are still likely to limit DHS authority and actions.

Then–Secretary Nielsen, in an editorial in the *Washington Post*, points out that the "U.S. government will remain unable to identify, track and mitigate weaponized or dangerous drones in our skies"[15] unless action is taken to change policy. Specifically, DHS lacks the authority and equipment to deal with UAS threats that are local to DHS assets. It is unable to intercept and override UAS control signals or adopt defensive actions, because current policy treats UAS as equivalent to manned aircraft.[16] Commandeering the control signal of a UAS also violates the Wiretap Act and Computer Fraud and Abuse Act, among other regulations.[17] The proposed legislation would give government agencies exemption to those laws specifically when dealing with UAS.[18] In addition, current law prohibits new defensive technologies to be tested in relevant operational environments such as urban areas and large public events that fall under DHS authority.

Whichever form future legislation takes, it will still restrict DHS actions and capabilities when combating a UAS threat subject to concerns for maintaining the privacy of citizens and freedom of the press to portray public opinions that can oppose DHS authority. Indeed, the recent Senate bill was quickly amended from its initial draft to include language that limits the reach and duration of DHS monitoring and countermeasures to "reasonable" levels. This proposed legislation suggests that technology enabling rapid detection, tracking, and identification of and response to threats will have high value to DHS. However, counter-UAS tools and concepts must limit collateral damage (e.g., hijacking a specific control signal rather than blanket jamming of broad bandwidth) and protect privacy.

---

[14] Nielsen, 2018.

[15] Nielsen, 2018.

[16] Nicholas Weaver, "The Necessary Authority to Counter Drone Threats," *Lawfare*, October 4, 2018.

[17] Weaver, 2018; United States Code, Title 18, Section 2511, *Interception and Disclosure of Wire, Oral. Or Electronic Communications Prohibited*, [date]; Public Law 99-474, The Computer Fraud and Abuse Act, 1986.

[18] U.S. Senate, Senate Resolution 115–2836, "Preventing Emerging Threats Act of 2018," May 14, 2018.

CHAPTER SIX

Conclusion and Recommendations

In this report, we focus on frameworks and approaches for understanding and documenting vulnerabilities and attack opportunities related to UAS and cybersecurity. In this chapter, a toolkit is presented that could help policymakers understand the UAS-related cybersecurity threat space and conduct a survey of today's threats across a variety of sources. We also provide a summary of future trends that may change the threat space over time. Finally, we focus on the relevance of these topics to the Department of Homeland Security.

This report focuses on classifying UAS threats largely with respect to cybersecurity, and presents an approach that can aid in analyzing these threats with the goal of directing efforts on mitigation and defensive strategies. However, the prioritization and likelihood of consequences from the highlighted threats are not addressed. Upon identifying and understanding these threat types, decisionmakers will still need to understand, for each threat, the likelihood of an attack, the consequences of such an attack, and the opportunities that exist to either prevent or exploit such an attack. As a first step in protecting itself from UAS-related cyberattacks or successfully using UAS as cyber assets, DHS can use the approaches outlined in this report to understand the set of attack vectors and attack surfaces. This is a necessary step, but it is not sufficient for establishing a coherent UAS and cybersecurity plan for cyber defense or offensive cyber operations.

Recommendations

Upon gaining a better understanding of the threat space, **DHS must continue to work with senior policymakers, cybersecurity experts, and other government and law enforcement agencies to move toward a coherent UAS cyber strategy**. This work will involve inventorying and categorizing UAS platforms, understanding possible consequences of as well as mitigation options for UAS-related cyberattacks, and staying abreast of new technological developments that could change the threat space. DHS should invest in operating a UAS test range or ranges in collaboration with the private sector, national labs, and other government stakeholders such as the FAA. This step would help ensure industry compliance with safety and security protocols and aid interagency coordination. Greater risks will accompany the capture and control of a larger, heavier fixed-wing drone, capable of flying far and carrying a heavy payload, than the typical small quadcopter drone flown by a hobbyist in the local park. Attacks involving the theft of UAS operator data will be more consequential when the operator is engaged in more sensitive operations; for example, monitoring critical infrastructure or protecting previously threatened facilities.

DHS should also prioritize the most critical vulnerabilities and find ways to close attack vectors and protect attack surfaces. To understand mitigation options, DHS will need to monitor technological development in counter-UAS systems and experiment with emerging attack techniques and technologies. A coordinated and updateable system of monitoring and intervention is likely to be required, as the innovation cycle of cyberattack and countermeasure suggests that even hardened systems cannot be guaranteed immune to attack.

Finally, **DHS will need to monitor UAS adoption and anticipate the implications of widespread UAS diffusion**. Capabilities such as autonomous flight and swarming will widen the UAS application space. As UAS are used in a wider range of activities, the number of legitimate-use UAS that are airborne at any given time will increase. From the perspective of threat mitigation, one of most important tasks in this new UAS-dense environment will be distinguishing licit from illicit activity.

Attack Categorization

In this appendix, we provide a taxonomy of types of attacks.

Table A.1
Attack Categorization

Attack Description	Attack Type*	UAS Role	Access Point	Bug/Weapon	Attack Vector	Date of Attack	Referenced in	Source Type
Drone hacking drone demonstration	D	Attacker	WiFi Network	Data Packet	Drone	2013	Kim Hartman and Keir Giles, "UAV Exploitation: A New Domain for Cyber Power," 2016 8th International Conference on Cyber Conflict (CyCon), Tallinn, Estonia: NATO Cooperative Cyber Defence Centre of Excellence, 2016.	Academic[a]
DEF CON Lecture on how to outfit a UAV and utilize wireless survey devices	I	Attacker	WiFi Network	Data Packet	Drone	2013		Academic[b]
Researchers develop app to turn drone into spying machine	S	Attacker	WiFi Network	Data Packet	Drone	2014	Chaitanya Rani, Hamidreza Modares, Raghavendra Sriram, Dariusz Mikulski, and Frank L. Lewis, "Security of Unmanned Aerial Vehicle Systems Against Cyber-Physical Attacks," Journal of Defense Modeling and Simulation: Applications, Methodology, Technology, Vol. 13, No. 3, July 1, 2016.	Academic[c]
DEF CON Lecture on how drones are used to monitor/hack information	I	Attacker	WiFi Network	Data Packet	Drone	2015		Academic[d]
Paper discusses Wireshark hacking	I	Attacker	WiFi Network	Data Packet	Drone	2016	Rani et al., 2016.	Academic[e]
Paper discusses MITM hacking	S	Attacker	WiFi Network	Data Packet	Drone	2016	Rani et al., 2016.	Academic[f]

Attack Description	Attack Type*	UAS Role	Access Point	Bug/Weapon	Attack Vector	Date of Attack	Referenced in	Source Type
Paper discusses Trojan Horse hacking	T	Attacker	WiFi Network	Data Packet	Drone	2016	Rani et al., 2016.	Academic[g]
DEF CON Lecture on drone defense products	T	Attacker	Drone RC Receiver	Data Packet	Drone	2017		Academic[h]
Drones can be used to "sniff" for unsecured wireless signals	I	Attacker	WiFi Network	Data Packet	Drone	2018	İsmail Güvenç, Farshad Koohifar, Simran Singh, Mihail L. Sichitiu, and David Matolak, "Amateur Drone Surveillance: Applications, Architectures, Enabling Technologies, and Public Safety Issues: Detection, Tracking, and Interdiction for Amateur Drones," IEEE Communications Magazine, April 2018.	Academic[i]
D13 releases technology to land enemy drones without jamming	E	Target	Drone RC Receiver	Data Packet	Anti-drone System	2016		Tech blog[j]
Elbit Systems releases ReDrone counter-UAV system	D	Target	Drone RC Receiver	Radio frequency	Anti-drone System	2016	Don Galeom, "As Drones Become Tools of War, Companies Turn to Hacking Them," Futurism, February 20, 2018.	Commercial[k]
Israeli Aerospace Industries release Drone Guard counter-UAV system	D	Target	Drone RC Receiver	Radio frequency	Anti-drone System	2016		Commercial[l]
SelexES releases Falcon Shield counter-UAV system	D	Target	Drone RC Receiver	Radio frequency	Anti-drone System	2016	Galeom, 2018.	Commercial[m]
Technology used to hunt down drones	I	Target	Data transmission	Data Packet	Anti-drone System	2018	Galeom, 2018.	News[n]

Attack Description	Attack Type*	UAS Role	Access Point	Bug/Weapon	Attack Vector	Date of Attack	Referenced in	Source Type
Apollo Shield releases counter-UAV system	D	Target	Drone RC Receiver	Radio frequency	Anti-drone System	2018	Galeom, 2018.	Commercial[o]
DEF CON Lecture on UAV applications and vulnerabilities	S	Target	WiFi Network	Data Packet	Cell Phone	2016		Academic[p]
Russians use white noise broadcasting to defend against Ukrainian UAVs	D	Target	Drone RC Receiver	Radio frequency	Jamming device	2014	Hartman and Giles, 2016.	Academic[q]
Researchers use shoulder-mounted jammer to attack drones	D	Target	Drone RC Receiver	Radio frequency	Jamming device	2015	C. G. Leela Krishna and Robin R. Murphy, "A Review on Cybersecurity Vulnerabilities for Unmanned Aerial Vehicles," 2017 IEEE International Symposium on Safety, Security and Rescue Robotics, Shanghai, China: IEEE, October 2017.	Academic[r]
DEF CON Lecture on drone defense products	D	Target	Drone GPS	Radio frequency	Jamming device	2017		Academic[s]
DEF CON Lecture on drone defense products	D	Target	Drone RC Receiver	Radio frequency	Jamming device	2017		Academic[t]
Pakistan seeks drone gun from China	D	Target	Drone RC Receiver	Radio frequency	Jamming device	2018		News[u]
Hezbollah claims to hack Israeli drone feed	I	Target	Data transmission	Data Packet	Laptop	1997	Krishna and Murphy, 2017.	Academic[v]
Insurgents hack U.S. drone video feeds	I	Target	Data transmission	Data Packet	Laptop	2009	Krishna and Murphy, 2017.	Academic[w]

Attack Description	Attack Type*	UAS Role	Access Point	Bug/Weapon Type	Attack Vector	Date of Attack	Referenced in	Source Type
Paper discusses GPS spoofing attack on UAVs	S	Target	Drone GPS	Radio frequency	Laptop	2013	Sait Murat Giray, "Anatomy of Unmanned Aerial Vehicle Hijacking with Signal Spoofing," *2013 6th International Conference on Recent Advances in Space Technologies (RAST)*, Istanbul, Turkey: IEEE, 2013.	Academic
Researchers deactivate drone with push of a button	D	Target	WiFi Network	Data Packet	Laptop	2015	Krishna and Murphy, 2017.	Academic[x]
DEF CON Lecture on hacking UAVs	S	Target	Drone GPS	Radio frequency	Laptop	2015		Academic[y]
Johns Hopkins researchers demonstrate multiple drone hacks	S	Target	WiFi Network	Data Packet	Laptop	2016		Academic[z]
Researcher demonstrates hacking of $35,000 police drone	S	Target	Drone RC Receiver	Radio frequency	Laptop	2016		Tech blog[aa]
Texas researchers demonstrate hacking of commercial UAV	T	Target	WiFi Network	Data Packet	Laptop	2017		News[bb]
Drones that utilize open WiFi communications are susceptible to hacks	S	Target	WiFi Network	Data Packet	Laptop	2018	Güvenç et al., 2018.	Academic[cc]
Demonstration that GPS can be hacked	S	Target	Drone GPS	Radio frequency	Spoofing device	2002	Giray, 2013.	Academic[dd]

Attack Description	Attack Type*	UAS Role	Access Point	Bug/Weapon	Attack Vector	Date of Attack	Referenced in	Source Type
Iran captures CIA surveillance drone	S	Target	Drone GPS	Radio frequency	Spoofing device	2011	Ahmad Y. Javaid, Farha Jahan, and Weiqing Sun, "Analysis of Global Positioning System-Based Attacks and a Novel Global Positioning System Spoofing Detection/Mitigation Algorithm for Unmanned Aerial Vehicle Simulation," Simulation: Transactions of the Society for Modeling and Simulation International, Vol. 93, No. 5, 2017.	Academic[ee]
Spoofing demonstration on civilian UAV	S	Target	Drone GPS	Radio frequency	Spoofing device	2012	Javaid, Jahan, and Sun, 2017.	Academic[ff]
Students hijack luxury yacht	S	Target	Drone GPS	Radio frequency	Spoofing device	2013	Javaid, Jahan, and Sun, 2017.	Academic[gg]
GPS jamming simulation	D	Target	Drone GPS	Radio frequency	Spoofing device	2017	Javaid, Jahan, and Sun, 2017.	Academic[hh]
GPS spoofing simulation	S	Target	Drone GPS	Radio frequency	Spoofing device	2017	Javaid, Jahan, and Sun, 2017.	Academic[ii]
Tamil rebels hijack U.S. satellite signal	E	Target	Unspecified	Data Packet	Unspecified	2007	Giray, 2013.	Academic[jj]
Computer virus logs pilot behavior during missions	I	Target	Unspecified	Data Packet	Unspecified	2011	Krishna and Murphy, 2017.	Academic[kk]
S-100 Camcopter gets GPS jammed	D	Target	Drone GPS	Radio frequency	Unspecified	2012	Krishna and Murphy, 2017.	Academic[ll]
User loses control of drone, causing collision with civilian	D	Target	WiFi Network	Data Packet	Unspecified	2014	Rani et al., 2016.	Academic[mm]

Attack Description	Attack Type*	UAS Role	Access Point	Bug/Weapon	Attack Vector	Date of Attack	Referenced in	Source Type
Russians use GPS spoofing to defend against Ukrainian UAVs	S	Target	Drone GPS	Radio frequency	Unspecified	2014	Hartman and Giles, 2016.	Academic[nn]
Paper discusses DDoS hacking of UAVs	D	Target	Drone RC Receiver	Data Packet	Unspecified	2015	Ahmad Javaid, Weiqing Sun, and Mansoor Alam, "Single and Multiple UAV Cyber-Attack Simulation and Performance Evaluation," *EAI Endorsed Transactions on Scalable Information Systems*, Vol. 2, No. 4, 2015.	Academic[oo]
Paper discusses multiple-target jamming attack	D	Target	Drone RC Receiver	Radio frequency	Unspecified	2015	Javaid, Sun, and Alam, 2015.	Academic[pp]
Paper discusses DDoS hacking	D	Target	Drone RC Receiver	Data Packet	Unspecified	2016	Rani et al., 2016.	Academic[qq]
Communication path deception attacks	S	Target	WiFi Network	Data Packet	Unspecified	2017	Lebsework Negash, Sang-Hyeon Kim, and Han-Lim Choi, "Distributed Observes for Cyberattack Detection and Isolation in Formation-Flying Unmanned Aerial Vehicles," *Journal of Aerospace Information Systems*, Vol. 14, No. 10, 2017.	Academic[rr]
Federal Trade Commission demonstrates multiple drone hacks	S	Target	WiFi Network	Data Packet	Unspecified	2017		Tech blog[ss]
Node attacks on UAV systems	D	Target	Unspecified	Radio frequency	Unspecified	2017	Negash, Kim, and Choi, 2017.	Academic[tt]

* D = Denial of Service, I = Information Disclosure, S = Spoofing, E = Elevation of Privilege, T = Tampering.

[a] Hak5, "Drones Hacking Drones (Part 1), Hak5 1518.1," video, YouTube, December 18, 2013.

[b] DEFCONConference, "DEF CON 21 – Ricky Hill - Phantom Network Surveillance UAV Drone," video, YouTube, December 23, 2013.

[c] Pierluigi Paganini, "Snoopy Software Can Turn A Drone is A Data Stealer," *SecurityAffairs*, March 24, 2014.

[d] DEFCONConference, "DEF CON 22 – Glenn Wilkinson - Practical Aerial Hacking and Surveillance," video, YouTube, January 6, 2015a.

[e] Chaitanya Rani, Hamidreza Modares, Raghavendra Sriram, Dariusz Mikulsku, and Frank L. Lewis, "Security of Unmanned Aerial Vehicle Systems Against Cyber-Physical Attacks," *Journal of Defense Modeling and Simulation: Applications, Methodology, Technology,* Vol. 13, No. 3, July 1, 2016.

[f] Rani et al., 2016.

[g] Rani et al., 2016.

[h] DEFCON Conference, "DEF CON 25 – Francis Brown, David Latimer - Putting the Emerging Drone Defense Market to the Test," video, YouTube, October 12, 2012.

[i] Jingjing Gu, Tao Su, Qiuhong Wang, Xiaojiang Du, and Mohsen Guizani, "Multiple Moving Targets Surveillance Based on a Cooperative Network for Multi-UAV," *IEEE Communications Magazine,* Vol. 56, No. 4, April 2018.

[j] DGIwire, "Landing Enemy Drones Safely: Here Comes the Technology," undated.

[k] "Elbit Systems Reveals ReDrone—An Advanced Anti-Drone Protection and Neutralization System," Elbit Systems, November 15, 2016.

[l] Israel Aerospace Industries, "ELI-4030 - Drone Guard - Lightweight Drone Detection, Identification and Disruption System," website, undated.

[m] SelexES, "Falcon Shield Counter-UAV System," 2017.

[n] Samuel Burke, "New Technology Created to Hunt Down Drones," *CNN*, August 8, 2018.

[o] "ApolloShield Counter-Drone Systems," webpage, undated.

[p] DEFCONConference, "DEF CON 24 Conference – Aaron Luo - Drones Hijacking: Multidimensional Attack Vectors and Countermea[sures]," video, YouTube, November 10, 2016

[q] Hartman and Giles, 2016.

[r] Matt Terndrup, "Long-Distance Jammer Is Taking Down Drones," *Make;* October 16, 2015.

[s] DEFCONConference, 2012.

[t] DEFCONConference, 2012.

[u] "Pakistan Seeks Drone Gun from China," *ANI,* June 8, 2018.

[v] Greg Grant, "Hezbollah Claims It Hacked Israeli Drone Video Feeds (Updated)," *Military.com,* August 10, 2018.

[w] Siobhan Gorman, Yochi J. Dreazen, and August Cole, "Insurgents Hack U.S. Drones," *Wall Street Journal*, December 17, 2009.

[x] Sean Gallagher, "Parrot Drones Easily Taken Down or Hijacked, Researchers Demonstrate: Open Telnet Port, Open Wi-Fi, Root Access, Open Season," *ArsTechnica,* August 15, 2015.

[y] DEFCONConference, "DEF CON 23 - Knocking My Neighbors Kids Cruddy Drone Offline," video, YouTube, December 25, 2015b.

[z] Phil Sneiderman, "Johns Hopkins Scientists Show How Easy It Is to Hack a Drone and Crash It," Johns Hopkins University, June 8, 2016.

[aa] Andy Greenberg, "Hacker Says He Can Hijack a $35k Police Drone A Mile Away," *Wired,* March 2, 2018.

[bb] Thomas Brewster, "Watch A Very Vulnerable $140 Quadcopter Drone Get Hacked Out Of The Sky," *Forbes,* April 25, 2017.

[cc] Güvenç et al., 2018.

[dd] Jon S. Warner and Roger G. Johnston, "A Simple Demonstration that the Global Positioning System (GPS) Is Vulnerable to Spoofing," *Journal of*

ee Daniel Shepard, Jahshan A. Bhatti, and Todd E. Humphreys, "Drone Hack: Spoofing Attack Demonstration on a Civilian Unmanned Aerial Vehicle," *GPS World*, August 1, 2012.

ff Shepard, Bhatti, and Humphreys, 2012.

gg Juha Saarinen, "Students Hijack Luxury Yacht with GPS Spoofing: No Alarms Triggered," *itnews*, July 30, 2013.

hh Javaid, Jahan, and Sun, 2017.

ii Javaid, Jahan, and Sun, 2017.

jj Stephen Northcutt, "Security Laboratory: Methods of Attack Series: Are Satellites Vulnerable to Hackers?" SANS Technology Institute, undated.

kk Noah Shachtman, "Exclusive: Computer Virus Hits U.S. Drone Fleet," *Wired*, October 7, 2011.

ll Krishna and Murphy, 2017.

mm Sean Gallagher, "Triathlete Injured by 'Hacked' Camera Drone," *ArsTechnica*, April 7, 2014.

nn Hartman and Giles, 2016.

oo Javaid, Sun, and Alam, 2015.

pp Xin Su, Shichao Yu, Jie Zeng, Yujun Kuang, Nayan Fang, and Zejiao Li, "Hierarchical Codebook Design for Massive MIMO," *2013 8th International Conference on Communications and Networking in China*, Guilin, China: IEEE, August 2013.

qq Rani et al., 2015.

rr Negash, Kim, and Choi, 2017, pp. 551–565.

ss April Glaser, "The U.S. Government Showed Just How Easy It Is to Hack Drones Made by Parrot, DBPower and Cheerson," *Recode*, January 4, 2017.

tt Negash, Kim, and Choi, 2017.

References

Abbaspour, Alireza, Michael Sanchez, Armen Sargolzaei, Kang Yen, and Nalat Sornkhampan, "Adaptive Neural Network Based Fault Detection Design for Unmanned Quadrotor Under Faults and Cyber Attacks," *2017 25th International Conference on Systems Engineering*, Las Vegas, Nev.: IEEE, 2017.

AirMap, *Five Critical Enablers for Safe, Efficient, and Viable UAS Traffic Management (UTM)*, white paper, January 2018.

Airobotics, "Automated Industrial Drones," webpage, undated. As of December 11, 2018: https://www.airoboticsdrones.com

Arkin, Ronald, *Behavior-Based Robotics*, Cambridge, Mass.: MIT Press, 1998.

Aktaş, Ziya A., Cemal Gemci, and Eray Yağdereli, "A Study on Cyber-Security of Autonomous and Unmanned Vehicles," *Journal of Defense Modeling and Simulation*, Vol. 12, No. 4, October 2015, pp. 369–381.

"ApolloShield Counter-Drone Systems," webpage, undated. As of December 11, 2017: https://www.apolloshield.com/

Applied Blockchain, "Applied Blockchain: Building Future-Proof Blockchain Applications," webpage, 2018. As of December 11, 2018: https://appliedblockchain.com/projects/

Association for Unmanned Vehicle Systems International, "Unmanned Systems and Robotics Database," webpage, undated. As of July 3, 2018: http://roboticsdatabase.auvsi.org/home?CLK=05da284f-5498-49d9-a548-91169efa9d65

Axe, David, "How Russia Says It Swatted Down a Drone Swarm in Syria," *Motherboard*, January 12, 2018. As of February 5, 20019: https://motherboard.vice.com/en_us/article/43qbbw/russia-says-it-swatted-down-drone-swarm-syria-isis

Barooah, Prabir, Gaemus E. Collins, and João P. Hespanha, "GeoTrack: Bio-Inspired Global Video Tracking by Networks of Unmanned Aircraft Systems," *Proceedings of the SPIE*, Vol. 7321, May 2009.

Belikovetsky, Sofia, Mark Yampolskiy, Jinghui Toh, Jacob Gatlin, and Yuval Elovici, "dr0wned– Cyber–Physical Attack with Additive Manufacturing," *WOOT '17 Proceedings of the 11th USENIX Conference on Offensive Technologies*, Vancouver, Canada, August 2017. As of February 5, 2019: https://itrust.sutd.edu.sg/research/projects/security-additive-manufacturing/

Bennett, Tim, "Air Based Technologies," presentation at the ATCA Aviation Cybersecurity Conference, 2017.

Blockchain, "About," webpage, 2019. As of December 11, 2012:
https://www.blockchain.com/about

Brewster, Thomas, "Watch A Very Vulnerable $140 Quadcopter Drone Get Hacked Out Of The Sky," *Forbes*, April 25, 2017. As of February 5, 2019:
https://www.forbes.com/sites/thomasbrewster/2017/04/25/
vulnerable-quadcopter-drone-hacked-by-ut-dallas-cyber-researchers/

Buczak, Anna L., and Erhan Guven, "A Survey of Data Mining and Machine Learning Methods for Cyber Security Intrusion Detection," *IEEE Communications Surveys & Tutorials*, Vol. 18, No. 2, Second Quarter, 2016, pp. 1153–1176.

Burke, Samuel, "New Technology Created to Hunt Down Drones," *CNN*, August 8, 2018. As of February 5, 2019:
https://www.cnn.com/videos/world/2018/08/08/bringing-drones-down-burke-pkg-vpx.cnn

Chao, Hsun, Apoorv Maheshwari, Varun Sudarsanan, Shashank Tamaskar, and Daniel DeLaurentis, "UAV Traffic Information Exchange Network," *2018 Aviation Technology, Integration, and Operations Conference*, Atlanta, Ga.: AIAA Aviation Forum, 2018.

Cichonski, Jeffrey, and Joshua Franklin, "LTE Security—How Good Is It?" presentation given at the 2015 RSA Conference, San Francisco, Calif., 2015. As of February 5, 2019:
https://www.rsaconference.com/writable/presentations/file_upload/tech-r03_lte-security-how-good-is-it.pdf

Clarivate Analytics, "Web of Science," webpage, undated. As of June 18, 2019:
https://clarivate.com/products/web-of-science/

Clarivate Analytics, "Derwent Innovations Index," database, 2019.

DEFCONConference, "DEF CON 25 – Francis Brown, David Latimer - Putting the Emerging Drone Defense Market to the Test," video, YouTube, October 12, 2012. As of February 5, 2019:
https://www.youtube.com/watch?v=LXFnyDihdl0

———, "DEF CON 21 – Ricky Hill - Phantom Network Surveillance UAV Drone," video, YouTube, December 23, 2013. As of February 5, 2019:
https://www.youtube.com/watch?v=bLOAdeWJYGg

———, "DEF CON 22 – Glenn Wilkinson - Practical Aerial Hacking and Surveillance," video, YouTube, January 6, 2015a. As of February 5, 2019:
https://www.youtube.com/watch?v=knrvrR-B1ZI

———, "DEF CON 23 - Knocking My Neighbors Kids Cruddy Drone Offline," video, YouTube, December 25, 2015b. As of February 5, 2019:
https://www.youtube.com/watch?v=5CzURm7OpAA&t=1439s

———, "DEF CON 24 Conference – Aaron Luo - Drones Hijacking: Multidimensional Attack Vectors and Countermea[sures]," video, YouTube, November 10, 2016. As of February 5, 2019:
https://www.youtube.com/watch?v=R6RZ5KqSVcg

DGIwire, "Landing Enemy Drones Safely: Here Comes the Technology," undated. As of December 11, 2018:
https://www.dgiwire.com/landing-enemy-drones-safely-here-comes-the-technology/?utm_
campaign=shareaholic&utm_medium=email_this&utm_source=email

DHS—*See* U.S. Department of Homeland Security.

Dorado, "Dorado ICO," webpage, undated. As of December 11, 2018:
https://www.dorado.tech

"Drone Use Reaches 'Landmark Level' in Harvey Disaster Response," *InfoGram*, Vol. 17, No. 37, September 14, 2017. As of May 29, 2019:
https://www.hsdl.org/?abstract&did=804423

"Elbit Systems Reveals ReDrone—An Advanced Anti-Drone Protection and Neutralization System," Elbit Systems, November 15, 2016. As of December 11, 2018:
http://elbitsystems.com/pr-new/elbit-systems-reveals-redrone-advanced-anti-drone-protection-neutralization-system/

Federal Aviation Administration, "System Wide Information Management (SWIM)," webpage, 2018. As of February 5, 2019:
https://www.faa.gov/air_traffic/technology/swim/

Galeom, Don, "As Drones Become Tools of War, Companies Turn to Hacking Them," *Futurism*, February 20, 2018. As of February 5, 2019:
https://futurism.com/drone-hack-technology/

Gallagher, Sean, "Triathlete Injured by 'Hacked' Camera Drone," *ArsTechnica*, April 7, 2014. As of February 5, 2019:
https://arstechnica.com/information-technology/2014/04/triathlete-injured-by-hacked-camera-drone/

———, "Parrot Drones Easily Taken Down or Hijacked, Researchers Demonstrate: Open Telnet Port, Open Wi-Fi, Root Access, Open Season," *ArsTechnica*, August 15, 2015. As of February 5, 2019:
https://arstechnica.com/information-technology/2015/08/parrot-drones-easily-taken-down-or-hijacked-researchers-demonstrate/

Ganjoo, Amit, "Blockchain and Drones—The Reality," ANRA Technologies, February 13, 2018. As of February 5, 2019:
http://www.anratechnologies.com/home/consulting/blockchain-and-drones-the-reality/

Gil-Casals, Silvia, Philippe Owezarski, and Gilles Descargues, "Generic and Autonomous System for Airborne Networks Cyber-Threat Detection, *2013 IEEE/AIAA 32nd Digital Avionics Systems Conference (DASC)*, October 2013.

Gil-Casals, Silvia, *Risk Assessment and Intrusion Detection for Airborne Networks*, dissertation, Toulouse, France: INSA Toulouse, 2014.

Giray, Sait Murat, "Anatomy of Unmanned Aerial Vehicle Hijacking with Signal Spoofing," *2013 6th International Conference on Recent Advances in Space Technologies (RAST)*, Istanbul, Turkey: IEEE, 2013.

Glaser, April, "The U.S. Government Showed Just How Easy It Is to Hack Drones Made by Parrot, DBPower and Cheerson," *Recode*, January 4, 2017. As of February 5, 2019:
https://www.recode.net/2017/1/4/14062654/drones-hacking-security-ftc-parrot-dbpower-cheerson

Gorman, Siobhan, Yochi J. Dreazen, and August Cole, "Insurgents Hack U.S. Drones," *Wall Street Journal*, December 17, 2009.

Grant, Greg, "Hezbollah Claims It Hacked Israeli Drone Video Feeds (Updated)," *Military.com*, August 10, 2018. As of February 5, 2019:
https://www.military.com/defensetech/2010/08/10/hezbollah-claims-it-hacked-israeli-drone-video-feeds

Greenberg, Andy, "Hacker Says He Can Hijack a $35k Police Drone A Mile Away," *Wired*, March 2, 2018. As of February 5, 2019:
https://www.wired.com/2016/03/hacker-says-can-hijack-35k-police-drone-mile-away/

Greig, Jonathan, "AI-Powered Autonomous Drone Could Bring New Capabilities to Agriculture, Logistics, More," *Tech Republic*, May 16, 2018. As of February 5, 2019:
https://www.techrepublic.com/article/scientists-create-miniature-drone-that-can-fly-itself-with-ai/

Güvenç, İsmail, Farshad Koohifar, Simran Singh, Mihail L. Sichitiu, and David Matolak, "Amateur Drone Surveillance: Applications, Architectures, Enabling Technologies, and Public Safety Issues: Detection, Tracking, and Interdiction for Amateur Drones," *IEEE Communications Magazine*, April 2018, pp. 75–81. As of February 5, 2019:
https://ieeexplore.ieee.org/stamp/stamp.jsp?tp=&arnumber=8337900

Hackett, Robert, "Wal-Mart Explores Blockchain for Delivery Drones," *Fortune*, May 30, 2017. As of February 5, 2019:
http://fortune.com/2017/05/30/walmart-blockchain-drones-patent/

Hackius, Niels, and Moritz Petersen, "Blockchain in Logistics and Supply Chain: Trick or Treat?" *Proceedings of the Hamburg International Conference of Logistics (HICL)*, Vol. 23, Hamburg, Germany: Hamburg International Conference of Logistics (HICL), 2017.

Hartman, Kim, and Keir Giles, "UAV Exploitation: A New Domain for Cyber Power," *2016 8th International Conference on Cyber Conflict (CyCon)*, Tallin, Estonia: NATO Cooperative Cyber Defence Centre of Excellence, 2016.

Hak5, "Drones Hacking Drones (Part 1), Hak5 1518.1," video, YouTube, December 18, 2013. As of February 5, 2019:
https://www.youtube.com/watch?time_continue=152&v=Fk1Bpy5ccPU

Hastie, Trevor, Robert Tibshirani, and Jerome Friedman, *The Elements of Statistical Learning*, New York: Springer, 2001.

Higgins, Fiona, Allan Tomlinson, and Keith M. Martin, "Survey on Security Challenges for Swarm Robotics," *2009 5th International Conference on Autonomic and Autonomous Systems*, Valencia, Spain: ICAS, April 2009.

Huber, Mark, "Study: Half of Drone Flights to Be Autonomous by 2022," *AIN Online*, March 22, 2018. As of February 5, 2019:
https://www.ainonline.com/aviation-news/business-aviation/2018-03-22/study-half-drone-flights-be-autonomous-2022

Iansiti, Marco, and Karim R. Lakhani, "The Truth About Blockchain," *Harvard Business Review*, January–February 2017, pp. 118–127.

Israel Aerospace Industries, "ELI-4030 - Drone Guard - Lightweight Drone Detection, Identification and Disruption System," webpage, undated. As of December 11, 2018:
https://www.iai.co.il/p/eli-4030-drone-guard

Javaid, Ahmad Y., Farha Jahan, and Weiqing Sun, "Analysis of Global Positioning System-Based Attacks and a Novel Global Positioning System Spoofing Detection/Mitigation Algorithm for Unmanned Aerial Vehicle Simulation," *Simulation: Transactions of the Society for Modeling and Simulation International*, Vol. 93, No. 5, 2017, pp. 427–441.

Javaid, Ahmad, Weiqing Sun, and Mansoor Alam, "Single and Multiple UAV Cyber-Attack Simulation and Performance Evaluation," *EAI Endorsed Transactions on Scalable Information Systems*, Vol. 2, No. 4, 2015.

Jingjing Gu, Tao Su, Qiuhong Wang, Xiaojiang Du, and Mohsen Guizani, "Multiple Moving Targets Surveillance Based on a Cooperative Network for Multi-UAV," *IEEE Communications Magazine*, Vol. 56, No. 4, April 2018, pp. 82–89.

Jones, Andrew, and Jeremy Straub, "Using Deep Learning to Detect Network Intrusions and Malware in Autonomous Robots," *Proceedings Vol. 10185, SPIE Defense + Security, Cyber Sensing*, April 2017.

Joshi, Divya, "Commercial Unmanned Aerial Vehicle (UAV) Market Analysis – Industry Trends, Companies and What You Should Know," *Business Insider*, August 8, 2017. As of August 24, 2018: http://www.businessinsider.com/commercial-uav-market-analysis-2017-8

Kamkar, Samy, "SkyJack," webpage, December 2, 2013. As of August 24, 2918: http://www.samy.pl/skyjack/

Kerns, Andrew J., Daniel P. Shepard, Jahshan A. Bhatti, and Todd E. Humphreys, "Unmanned Aircraft Capture and Control Via GPS Spoofing," *Journal of Field Robotics*, Vol. 31, No. 4, July/August 2014, pp. 617–636.

Khandelwal, Swati, "MalDrone—First Ever Backdoor Malware for Drones," *Hacker News*, January 27, 2015. As of February 5, 2019: https://thehackernews.com/2015/01/MalDrone-backdoor-drone-malware.html

Kopardekar, Parimal, Joseph Rios, Thomas Prevot, Marcus Johnson, Jaewoo Jung, and John E. Robinson, III, "Unmanned Aircraft System Traffic Management (UTM) Concept of Operations," *16th AIAA Aviation Technology, Integration, and Operations Conference*, Washington, D.C.: AIAA Aviation, 2016.

Kuhn, Kenneth, *Small Unmanned Aerial System Certification and Traffic Management Systems*, Santa Monica, Calif.: RAND Corporation, PE-269-RC, 2017. As of February 5, 2019: https://www.rand.org/pubs/perspectives/PE269.html

Krishna, C. G. Leela, and Robin R. Murphy, "A Review on Cybersecurity Vulnerabilities for Unmanned Aerial Vehicles," *2017 IEEE International Symposium on Safety, Security and Rescue Robotics*, Shanghai, China: IEEE, October 2017.

"Leading the World to 5G: Evolving Cellular Technologies for Safer Drone Operation," Qualcomm, September 6, 2016. As of February 5, 2019: https://www.qualcomm.com/media/documents/files/leading-the-world-to-5g-evolving-cellular-technologies-for-safer-drone-operation.pdf

Malhotra, Raj P., Michael J. Pribilski, Patrick A. Toole, and Craig Agate, "Decentralized Asset Management for Collaborative Sensing," *Proceedings Volume 10194, Micro- and Nanotechnology Sensors, Systems, and Applications IX*, Anaheim, Calif.: SPIE Defense + Security, May 18, 2017.

Margaritoff, Marco, "MIT's NanoMap Tech Allows for Consistent, High-Speed, Autonomous Drone Navigation," *The Drive*, February 12, 2018. As of February 5, 2019: http://www.thedrive.com/aerial/18429/mits-nanomap-tech-allows-for-consistent-high-speed-autonomous-drone-navigation

———, "World's Smallest Autonomous Drone Takes Flight in Europe," *The Drive*, May 31, 2018. As of February 5, 2019: http://www.thedrive.com/tech/21203/the-worlds-smallest-autonomous-drone-takes-flight-in-europe

Mozur, Paul, "Drone Maker D.J.I. May Be Sending Data to China, U.S. Officials Say," *New York Times*, November 29, 2017.

Negash, Lebsework, Sang-Hyeon Kim, and Han-Lim Choi, "Distributed Observes for Cyberattack Detection and Isolation in Formation-Flying Unmanned Aerial Vehicles," *Journal of Aerospace Information Systems*, Vol. 14, No. 10, 2017, pp. 551–565.

Newman, Lily Hay, "Hacker Lexicon: What Is an Attack Surface," *Wired*, March 12, 2017. As of February 5, 2019:
https://www.wired.com/2017/03/hacker-lexicon-attack-surface/

Nielsen, Kirstjen M., "The U.S. Isn't Prepared for the Growing Threat of Drones," *Washington Post*, July 4, 2018.

Northcutt, Stephen, "Security Laboratory: Methods of Attack Series: Are Satellites Vulnerable to Hackers?" SANS Technology Institute, undated. As of February 5, 2019:
https://www.sans.edu/cyber-research/security-laboratory/article/satellite-dos

"OFFensive Swarm-Enabled Tactics (OFFSET)," Defense Advanced Research Projects Agency, undated. As of December 11, 2018:
https://www.darpa.mil/work-with-us/offensive-swarm-enabled-tactics

Osborne, Charlie, "Maldrone: Malware Which Hijacks Your Personal Drone," *ZDNet.com*, January 27, 2015. As of August 24, 2018:
https://www.zdnet.com/article/maldrone-malware-which-hijacks-your-personal-drones/

Otto, Robert P., "Small Unmanned Aircraft Systems (SUAS) Flight Plan: 2016-2036. Bridging the Gap Between Tactical and Strategic," Technical Report, Air Force Deputy Chief of Staff, Washington, D.C., April 30, 2016.

Paganini, Pierluigi, "Snoopy Software Can Turn A Drone is A Data Stealer," *SecurityAffairs*, March 24, 2014. As of February 5, 2019:
http://securityaffairs.co/wordpress/23374/hacking/snoopy-drone-data-stealer.html

"Pakistan Seeks Drone Gun from China," *ANI*, June 8, 2018. As of February 5, 2019:
https://www.msn.com/en-xl/asia/top-stories/pakistan-seeks-drone-gun-from-china/ar-BBLAvNi

Rani, Chaitanya, Hamidreza Modares, Raghavendra Sriram, Dariusz Mikulsku, and Frank L. Lewis, "Security of Unmanned Aerial Vehicle Systems Against Cyber-Physical Attacks," *Journal of Defense Modeling and Simulation: Applications, Methodology, Technology*, Vol. 13, No. 3, July 1, 2016, pp. 331–342.

Reed, Theodore, Joseph Geis, and Sven Dietrich, "SkyNET: A 3G-Enabled Mobile Attack Drone and Stealth Botmaster," *Proceedings of the 5th USENIX Conference on Offensive Technologies*, San Francisco, Calif.: WOOT '11, 2011.

Rios, Joseph, "Unmanned Aircraft System (UAS) Traffic Management (UTM)," National Aeronautics and Space Administration, February 15, 2019. As of December 11, 2018:
https://utm.arc.nasa.gov/index.shtml

Rogers, Everett M., *Diffusion of Innovations*, 5th ed., New York: Free Press (Simon and Schuster), 2003.

Ronen, Eyal, Adi Shamir, Achi-Or Weingarten, and Colin O'Flynn, "IoT Goes Nuclear: Creating a ZigBee Chain Reaction," *2017 IEEE Symposium on Security and Privacy*, San Jose, Calif.: IEEE, June 2017, pp. 195–212.

Saarinen, Juha, "Students Hijack Luxury Yacht with GPS Spoofing: No Alarms Triggered," *itnews*, July 30, 2013. As of February 5, 2019:
https://www.itnews.com.au/news/students-hijack-luxury-yacht-withgps-spoofing-351659

Sandia Technical Staff, interviews, 2010–2014.

Schmid, Jon, and Fei-Ling Wang, "Beyond National Innovation Systems: Incentives and China's Innovation Performance," *Journal of Contemporary China*, Vol. 26, No. 104, 2017, pp. 280–296.

Sedjelmaci, Hichem, and Sidi Mohamed Senouci, "Cyber Security Methods for Aerial Vehicle Networks: Taxonomy, Challenges and Solution," *Journal of Supercomputing*, Vol. 74, No. 10, October 2018, pp. 4928–4944.

SelexES, "Falcon Shield Counter-UAV System," 2017. As of December 11, 2018:
http://www.us.selex-es.com/-/falconshield

Sesar Joint Undertaking, "U-space," webpage, 2019. As of December 11, 2018:
https://www.sesarju.eu/U-Space

Shachtman, Noah, "Exclusive: Computer Virus Hits U.S. Drone Fleet," *Wired*, October 7, 2011. As of February 5, 2019:
https://www.wired.com/2011/10/virus-hits-drone-fleet/

Shakhatreh, Hazim, Ahmad Sawalmeh, Ala Al-Fuqaha, Zuochao Dou, Eyad Almaita, Issa Khalil, Noor Shamsiah Othman, Abdallah Khreishah, and Mohsen Guizani, "Unmanned Aerial Vehicles: A Survey on Civil Applications and Key Research Challenges," arXiv.org, arXiv:1805.00881, 2018.

Shepard, Daniel, Jahshan A. Bhatti, and Todd E. Humphreys, "Drone Hack: Spoofing Attack Demonstration on a Civilian Unmanned Aerial Vehicle," *GPS World*, August 1, 2012. As of February 5, 2019:
http://gpsworld.com/drone-hack/

Shostack, Adam, *Threat Modeling: Designing for Security*, Hoboken, N.J.: Wiley, 2014.

Skydio, "The Self-Flying Camera Has Arrived," webpage, 2019. As of December 11, 2018:
https://www.skydio.com

Sneiderman, Phil, "Johns Hopkins Scientists Show How Easy It Is to Hack a Drone and Crash It," Johns Hopkins University, June 8, 2016. As of August 24, 2018:
https://hub.jhu.edu/2016/06/08/hacking-drones-security-flaws/

Statt, Nick, "Skydio's AI-Powered Autonomous R1 Drone Follows You Around in 4K," *TheVerge*, February 13, 2018. As of February 5, 2019:
https://www.theverge.com/2018/2/13/17006010/
skydio-r1-autonomous-drone-4k-video-recording-ai-computer-vision-mapping

Su, Xin, Shichao Yu, Jie Zeng, Yujun Kuang, Nayan Fang, and Zejiao Li, "Hierarchical Codebook Design for Massive MIMO," *2013 8th International Conference on Communications and Networking in China*, Guilin, China: IEEE, August 2013.

"Technology: Drones," University of Illinois Extension, 2019. As of February 5, 2019:
https://4h.extension.illinois.edu/members/projects/technology-drones

Thales, "Thales Launches Ecosystem UTM and Joins Forces with Unifly to Facilitate Drone Use," July 3, 2017. As of February 5, 2019:
https://www.thalesgroup.com/en/worldwide/aerospace/press-release/
thales-launches-ecosystem-utm-and-joins-forces-unifly-facilitate

Terndrup, Matt, "Long-Distance Jammer Is Taking Down Drones," *Make:*, October 16, 2015. As of As of February 5, 2019:
https://makezine.com/2015/10/16/research-company-takes-aim-uavs-portable-anti-drone-

"The Cyber Killchain Framework," webpage, Lockheed Martin, 2019. As of February 5, 2019:
https://lockheedmartin.com/en-us/capabilities/cyber/cyber-kill-chain.html rifle/

Trujano, Fernando, Benjamin Chan, Greg Beams, and Reece Rivera, "Security Analysis of DJI Phantom 3 Standard," Massachusetts Institute of Technology, May 11, 2016. As of February 5, 2019:
https://courses.csail.mit.edu/6.857/2016/files/9.pdf

USCG—*See* U.S. Coast Guard.

U.S. Department of Homeland Security, "Program Executive Office for Unmanned Aerial Systems," webpage, undated. As of February 5, 2019:
https://www.dhs.gov/science-and-technology/peo-uas

———, "Unmanned Aircraft Systems (UAS) – Critical Infrastructure," webpage, undated. As of June 13, 2018:
https://www.dhs.gov/cisa/uas-critical-infrastructure

———, "Enabling Unmanned Aircraft Systems," DHS Science and Technology Directorate, May 3, 2017. As of February 5, 2019:
https://www.dhs.gov/sites/default/files/publications/Enabling%20UAS.%20Factsheet.pdf

U.S. Department of Homeland Security, Customs and Border Protection, "CBP Air and Marine Operations Conducting Third Deployment of UAS at San Angelo," February 27, 2018. As of June 13, 2018:
https://www.cbp.gov/newsroom/local-media-release/
cbp-air-and-marine-operations-conducting-third-deployment-uas-san

———, "CBP to Test the Operational Use of Small Unmanned Aircraft Systems in 3 U.S. Border Patrol Sectors," September 14, 2017. As of February 5, 2019:
www.cbp.gov/newsroom/national-media-release/
cbp-test-operational-use-small-unmanned-aircraft-systems-3-us-border

U.S. Coast Guard, "Unmanned Aircraft System," undated. As of June 13, 2018:
www.dcms.uscg.mil/Our-Organization/Assistant-Commandant-for-Acquisitions-CG-9/Programs/
Air-Programs/UAS

U.S. Federal Emergency Management Agency, "Transforming Cardiac Emergency Care with Drone Delivery of AEDs," webpage, 2018. As of February 5, 2019:
https://www.usfa.fema.gov/current_events/022218.html

U.S. Senate, Senate Resolution 115–2836, "Preventing Emerging Threats Act of 2018," May 14, 2018.

Valente, Junia, and Alvaro A. Cardenas, "Understanding Security Threats in Consumer Drones Through the Lens of the Discovery Quadcopter Family," *Proceedings of the 2017 Workshop on Internet of Things Security and Privacy*, Dallas, Tex.: Association for Computing Machinery, 2017, pp. 31–36.

Van Trees, Harry L., Kristine L. Bell, and Zhi Tian, *Detection, Estimation, and Modulation Theory: Part I–Detection, Estimation, and Filtering Theory*, 2nd ed., Hoboken, N.J.: Wiley, 2013.

Votiro Labs, "AutoCad Security Bug Still a Risk, It Turns Out," webpage, January 3, 2018. As of December 11, 2017:
https://www.votiro.com/autocad-security-bug-still-a-risk-it-turns-out/

Walters, Sander, "How Can Drones Be Hacked? The Updated List of Vulnerable Drones and Attack Tools," October 29, 2016. As of February 5, 2019:
https://medium.com/@swalters/how-can-drones-be-hacked-the-updated-list-of-vulnerable-drones-attack-tools-dd2e006d6809

Ward, Antonia, "Bitcoin and the Dark Web: The New Terrorist Threat?" The RAND Blog, January 22, 2018. As of December 11, 2018:
https://www.rand.org/blog/2018/01/bitcoin-and-the-dark-web-the-new-terrorist-threat.html

Warner, Jon S., and Roger G. Johnston, "A Simple Demonstration that the Global Positioning System (GPS) Is Vulnerable to Spoofing," *Journal of Security Administration*, Vol. 25, No. 2, 2002.

Weaver, Nicholas, "The Necessary Authority to Counter Drone Threats," *Lawfare*, October 4, 2018. As of May 29, 2019:
https://www.lawfareblog.com/necessary-authority-counter-drone-threats

Wierzbanowski, Scott, "Gremlins," Defense Advanced Research Projects Agency, undated. As of February 5, 2019:
https://www.darpa.mil/program/gremlins

Woodruff, Betsy, "ICE Wants Drones," *Daily Beast*, April 27, 2018. As of June 13, 2018:
https://www.thedailybeast.com/ice-wants-drones

Zaki, Maged, "Path to 5G: Building a Highway in the Sky for Autonomous Drones," Qualcomm Technologies, November 9, 2016. As of February 5, 2019:
https://www.qualcomm.com/news/onq/2016/11/09/
path-5g-building-highway-sky-autonomous-drones